T0135063

Studies in Systems, Decision and Control

Volume 138

Series editor

Janusz Kacprzyk, Polish Academy of Sciences, Warsaw, Poland
e-mail: kacprzyk@ibspan.waw.pl

The series "Studies in Systems, Decision and Control" (SSDC) covers both new developments and advances, as well as the state of the art, in the various areas of broadly perceived systems, decision making and control-quickly, up to date and with a high quality. The intent is to cover the theory, applications, and perspectives on the state of the art and future developments relevant to systems, decision making, control, complex processes and related areas, as embedded in the fields of engineering, computer science, physics, economics, social and life sciences, as well as the paradigms and methodologies behind them. The series contains monographs, textbooks, lecture notes and edited volumes in systems, decision making and control spanning the areas of Cyber-Physical Systems, Autonomous Systems, Sensor Networks, Control Systems, Energy Systems, Automotive Systems, Biological Systems, Vehicular Networking and Connected Vehicles, Aerospace Systems, Automation, Manufacturing, Smart Grids, Nonlinear Systems, Power Systems, Robotics, Social Systems, Economic Systems and other. Of particular value to both the contributors and the readership are the short publication timeframe and the world-wide distribution and exposure which enable both a wide and rapid dissemination of research output.

More information about this series at http://www.springer.com/series/13304

Mikuláš Hajduk · Marek Sukop
Matthias Haun

Cognitive Multi-agent Systems

Structures, Strategies and Applications to Mobile Robotics and Robosoccer

Mikuláš Hajduk
Department of Robotics
Technical University of Košice
Košice, Slovakia

Matthias Haun
Altrip, Germany

Marek Sukop
Department of Robotics
Technical University of Košice
Košice, Slovakia

ISSN 2198-4182 ISSN 2198-4190 (electronic)
Studies in Systems, Decision and Control
ISBN 978-3-030-06706-9 ISBN 978-3-319-93687-1 (eBook)
https://doi.org/10.1007/978-3-319-93687-1

Printed on acid-free paper

This Springer imprint is published by the registered company Springer International Publishing AG part of Springer Nature
The registered company address is: Gewerbestrasse 11, 6330 Cham, Switzerland

Preface

FIRA organization (Federation International Robot-soccer Association), a brainchild of Jong-Hwan Kima, began in 1995 in Korea. It has organized the world championships in various categories of robotic soccer since 1996.

Professor Mikuláš Hajduk and Marek Sukop visited AUSTRO TU Vienna in 2002, the most successful team in Europe at that time, developed by professor Kopacek. Their goal was to create robotic soccer team at SjF TUKE as well. The team SjF TUKE Robotics took part at European Championships in 2004 in German city Munich for the first time. Two years later, the team became the European champion and in 2010 the world champion.

The authors of the publication summarized their own knowledge and experience gained at the development and implementation of autonomous robotic system which consists of five mobile robot-soccer players. They concentrate on general description of the agents in multi-agent systems in the first chapters, by means of joining MAS and mobile robotics and robosoccer as an MAS test application in mobile robotics. Other chapters of the publications aim at the development and implementation process of the team. There are detailed descriptions of the system parts in some cases. In the end, there are two examples of the created strategies where methodology of creating strategies was used depending on the project of the main agent as a strategy owner.

Košice, Slovakia — Mikuláš Hajduk
Košice, Slovakia — Marek Sukop
Altrip, Germany — Matthias Haun

Contents

Summary

Robotic soccer is the application designed for trying out different approaches to the multi-robot system operation that prefers individual teams. International competitions take place every year in this category. Many scientists deal with them since it is a great platform for the study of intelligent operation and much knowledge might be used in industry with similar problems, e.g. military or service activities. This is the reason why the opinion is being spread around the world that the research in this field of study is meaningful and excellent topic even further. SjF TUKE Robotics has achieved fantastic results at international competitions for several years by this attitude, by means of above-mentioned approach towards MAS, its hierarchical arrangement and cooperation of agents within described strategies.

Introduction

Techniques of the artificial intelligence (AI) were applied on problem solving of the distributed computing problems by the distributed artificial intelligence (AI). A new field of study called multi-agent systems was developed in the 1990s focusing mainly on the behaviour management models as an eventual contrast to information management. Multi-agent systems provide solutions for complex problems, their subsystem decomposition which are distributed by independent solver—agent with its own interests and objectives [25].

Solutions developed by multi-agent approach to mobile robots control and tested on robosoccer will be applied on different groups of robotics teams with slight modifications in the future (lifeguard robots, rescue robots, firefighting robot, etc.).

Robosoccer is the game similar to a standard soccer; however, there is one significant difference: players are not people, but mobile robots. Robosoccer was created as an excellent application for testing AI and strategy development of group of cooperating mobile robots. Its significant feature is the fact that each team has to think in real time and be constantly ready and aware of dynamic changes of environment in which they operate. They must concentrate on achieving the main goal, i.e. score more goals than their opponent. It is thus necessary to observe the behaviour of opponents' robots, not only the ball. Multi-agent approach, also used in SjF TUKE robotics, is often employed to solve mobile robots' group control. There are many versions of competitions in robosoccer where mobile robots represent biped humanoids, quadruped robots—mostly robotic dogs, wheelbots, etc. Different simulation leagues also represent one of the versions. Category Mirosot is one the popular category of robosoccer where mini wheelbots (cube type) with the size of: 5 cm × 7.5 cm × 7.5 cm with maximum weight 600 g play soccer. Five robots play against another five (middle league, playground 220 cm × 180 cm), or 11 against 11 (large league playground 440 cm × 280 cm). The robot construction is significantly influenced by FIRA rules, and even though at mobile robots construction, different construction variants were tested. Electronic hardware and software offered plenty of ways how to bring this application into effect. There appeared to be many different ways and applications of various parts of AI regarding programming. In the world of above-mentioned category, there exist several research teams which check whether the solutions at the world champions are correct. Slovakia joined Asian countries in the last few years which have played the first fiddle since 2005.

Research trends in robosoccer include wide range of scientific areas such as construction principles, drive units, camera systems, new method implemented in AI and communication technologies, sophisticated software. Robosoccer became the significant scientific solution base at AI problem solving in the last few years, including robot construction, autonomous agents group, and image, operation and regulation technology processing.

Chapter 1
Multi-agent Systems—Terminology and Definitions

Multi-agent systems originated as an extension of the field of distributed artificial intelligence which is an intersection of two fields [25]:

– artificial intelligence
– distributed computing

The field of distributed computing has been in existence for such a long time as the solution of one computing problem might be divided into more processors. Originally these processors used data linked to problem solving and their fundamental goal was to facilitate their parallelism and synchronization [11].

Further extension of the distributed computing is closely related to the development of artificial intelligence which applied distribution approach on operation problems used by many processors. The field called distributed artificial intelligence appeared to be the result of this fusion. While distributed computing aims at low-level parallelism and synchronization, distributed artificial intelligence focuses on problem solving, communication and coordination. The distributed artificial intelligence was further divided into parallel artificial intelligence, distributed expert systems, even distributed knowledge sources and distributed problem solving due to the development of information management [11].

Distributed artificial intelligence may be further subdivided into [25]:

– distributed problem solving
– multi-agent systems: behavior coordination or behavior management. Usually no other agents are guaranteed. Mutual interactions of the agents are much more complex than behavior of the individuals.

M. Hajduk et al., *Cognitive Multi-agent Systems*, Studies in Systems, Decision and Control 138, https://doi.org/10.1007/978-3-319-93687-1_1

1.1 The Agent

There are many attempts to come up with a definition of an "agent" (in the field of multi-agent systems) that would best match the fundamentals of the agent in distributed artificial intelligence as it is with the defining the concept of "artificial intelligence", too. In spite of the fact that there does not exist any universal and established definition nowadays, one of the most popular definitions the one of Wooldridge and Jennings [53], which distinguishes so called strong notion of agent from the weak notion of agent. The weak agent is defined as:

Hardware or (more usually) software-based computer system that meets the following properties:

Autonomy agents operate without the direct intervention of humans or others, and have some kind of control over their actions and internal state;
Social ability agents interact with other agents (and possibly humans) via some kind of agent-communication language;

Reactivity agents perceive their environment, (which may be the physical world, a user via a graphical user interface, a collection of other agents, the Internet, or perhaps all of these combined), and respond in a timely fashion to changes that occur in it;

Pro-activity agents do not simply act in response to their environment, they are able to exhibit goal-directed behaviour by taking the initiative.

When identifying the *strong* notion of agent, in addition to having the properties identified above, concepts that are more usually applied to humans are implemented. For example, it is quite common in AI to characterize an agent using mentalistic or emotional notions, such as knowledge, belief, intention, etc.

Abovementioned approach was redefined in Nwan's work [53], where the definitions of an agent is used based on the following properties:

Autonomy an agent is able to operate independently without any human intervention even if the world an agent acts in is not easy to describe. A key element of autonomy is pro-activity, in other words the ability to take the initiative and act in a goal-oriented way rather than simply acting in response to the environment.

Cooperation the group of social skills which enables an agent to communicate with other agents or people by means of some communication language. Cooperation je one of the most essential properties of multi-agent systems, according to [35] are these agents able to coordinate their actions even without any communication.

Learning an agent needs to be capable of reaction in dynamic and undetermined environment and act in it. It is thus necessary to adopt learning factor by means of the interaction of an agent with the environment. It is assumed that learning may gradually improve behavior quality of an agent in the environment.

Properties that an agent should dispose of and express can be found in the contribution of Franklin and Graesser [15] where many other definitions are discussed in details. The following interpretations of the notion *agent* can be stated:

Maes [27]: An autonomous agent is the computational system that inhabits some complex dynamic sense and acts autonomously in this environment, and by doing so achieves a set of goals or tasks for which it is designed.

Hayes-Roth [23]: Intelligent agents continuously perform three functions: perception of dynamic conditions in the environment; affecting the environment by its own behaviour; and consideration (how to interpret perceptions, solve problems, draw inferences, and define actions.

Russel, Norvig [38]: "An agent" is anything that can be viewed as perceiving its environment through sensors and acts upon that environment through effectors.

Green, Hurst, Nagle, Cunningham, Somers, Evans [17]: "An agent is the computational entity which carry out some set of operations on behalf of another entity with some degree of pro-activity and/or reactivity, and in so doing, employs key properties of learning, cooperation and mobility".

From above mentioned definitions [15] and further assumptions there was created the concept of an agent which perceives an agent in the environment according to its capabilities to effect its future perception of the environment and its existence in time:

An autonomous agent is a system situated within an environment that perceives and acts on it, over time, in pursuit of its own agenda and so as to effect what it senses in the future.

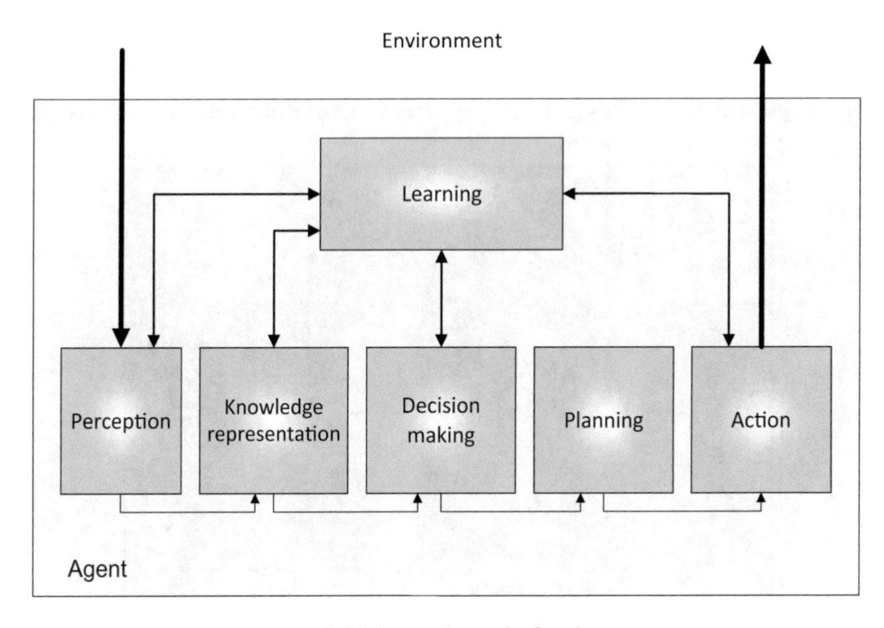

Fig. 1.1 An agent in typical UI, divided according to its functions

An architecture for planning systems of standard AI can be characterized by the decomposition in reference to functions. Each agent module is responsible for performing specific functions. Central system network of an agent decomposed by functions is in the Fig. 1.1 [26].

Decomposition of an agent based on activities (behavior) it performs is considered to be the alternative form. The system is divided into several parts (layers) whereas all the layers link perception to action in allocated context. Each layer is thus simple agent from one point of view and we might apprehend the whole system as multi-agent system. An agent with architecture scheme decomposed by means of its activities is in the Fig. 1.2 [26].

1.2 The Agent Types

Individual capability of agent to consider different solution approaches to achieve the goal puts an agent into four categories [25]:

- re-active agent
- model-based reflex (intentional) agent
- hybrid agent
- agent with behavior-based architecture

1.2.1 Reactive Agent

Reactive agent always perceives its environment, and responds in a timely fashion to changes that occur in it in order to satisfy its design objectives. It disposes of pre-defined set of activities (e.g. stereotype plans). The selection out of the actions

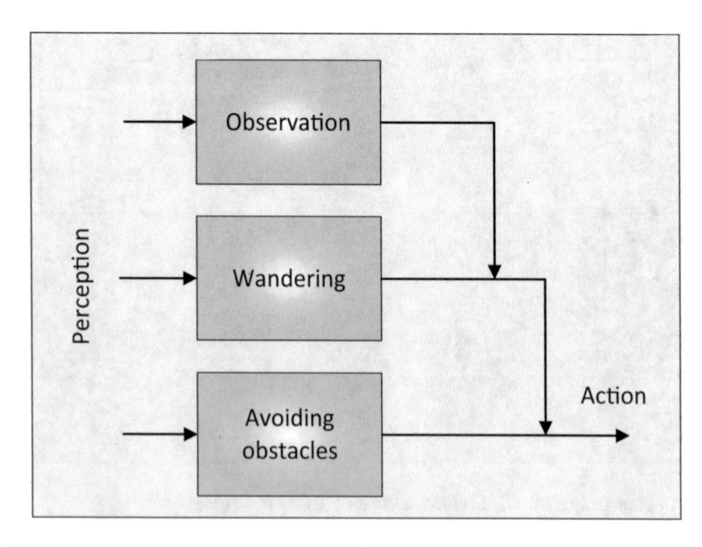

Fig. 1.2 An agent divided according to its acting

depends on goal-achievement whether the current world in the way it is perceived by an agent complies with the conditions required for activation of one of the actions. An agent modifies its inner environmental model in accordance with the stimuli from outer world. It is capable of informing other agents of all its results and findings. Providing that the stimuli from outer world keep repeating, cycle might occur.

Reactive agents are relatively simple, they can cooperate with other agents on low level. They can be characterized by the following properties:

Emergent functionality dynamic response to the condition of environment leading to surprise functionality. Its consequence seems to be the lack of accurate behavior specification of reactive agent (Behavior of reactive agent is directly dependent upon current set of outer circumstances).

Task decomposition reactive agent in fact represents the set of mutually independent modules that work on their own They are responsible for specific subtask; communication among modules is minimized and on a low level of abstraction; none of the agents dispose of global model of cooperating group, therefore not even global behavior accurately specified.

Manipulation with low-level data functions reactive agents perform with captured sensor information (This approach contrasts with high-level abstractions through symbols at deliberative agents).

Brooks [6], developed so called *subsumption* architecture [4] being applied on real mobile robots. Brooks' subsumption architecture can be described as a hierarchy of task-oriented models, i.e. fixed action patterns. Each level of hierarchy represented a behavior type of a particular complexity (e.g. avoiding obstacles, etc.). Final systems were eventually extremely simple without any explicit representation and consideration taking into account the difficulty and number of computing operations. However, Brooks demonstrated even such tasks performed by robots where it would be impressive if those tasks were solved by means of symbolic system.

Based on Brooks affirmations it is presumed that intelligent systems can be created out of simple agents without any inner symbolic models, while intelligent behavior of agents is achieved as a dynamic accordance of their own modules.

1.2.2 Model-Based Reflex (Intentional) Agent

Model-bask, led reflex agent considers its options to achieve its goal. Developing a plan and agent's motivation are closely related. Action coordination of group of agents is on high level, agents inform each other of their plans. These architectures use centralized inner module of the world. Its function is to interpret available sensor information and derivation of appropriate action. An action, or sequence of actions is normally produced by plan system that looks for possible sequences "condition $\rightarrow$ action" in the inner world module. Figure 1.3 displays possible bonds of processors in deliberative agent [26].

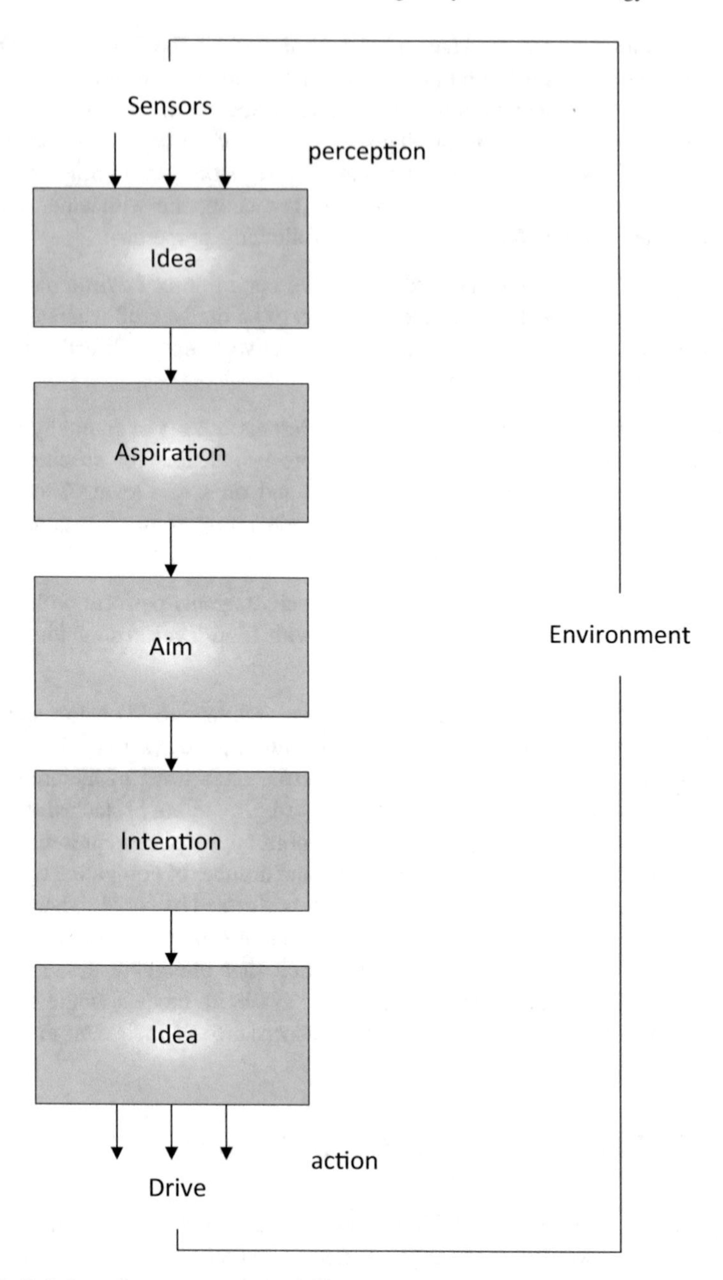

Fig. 1.3 Relations among processes in deliberate agent

Figure 1.3 Bonds between processes in model-based agent.

Creating actions and plans is much more time-consuming than selection of reactions at reactive agent. It is also necessary that all kinds of vital information are available in the inner world module. Current condition will be derived from it.

Performance of an agent requires problem-solving of transformation of real world into its symbolic representation in adequate time and problem-solving of symbolic representation of complex information about entities of the world and processes. It is also necessary to find the way how to deal with the information in order to achieve the results meeting at least time requirements within this consideration.

In case of mobile robots in real environment, a typical noise level and inaccuracy of sensory data in constantly changing dynamic environment. Dynamic and often incomplete real environment from point of view of read data asks for re-planning of current sequence of actions because of its continuous changing and it leads to increase of time complexity needed for adequate action deduction. Deliberate architectures were often criticized for their insufficient ability to work in real environment with high work complexity because they are unable to generate adequate actions in real time [5].

1.2.3 Hybrid Agent

Hybrid agent combines both above mentioned perspectives. Selection of properties starts only after all the positive features of reactive and model-based reflex agents have been considered in such a way so that hybrid agent satisfies the condition of reactivity response and rational decisions of deliberative part in real time. Agent consists of two independent parts:

- Reactive subcomponent would be capable to respond to world changes without any complex reasoning and decision-making in a short time period, while decision-making proceeds only on the sensor data level (e.g. instant obstacle avoidance manoeuvre, etc.).
- Deliberative subcomponent would be responsible for abstract planning and decision-making using symbolic representations of the world operating in longer time periods (e.g. generation of plans to achieve long-term objectives etc.).

Special interface design between both parts is typically the biggest problem when creating hybrid agents. The biggest priority with reference to actions selection should have the reactive subcomponent which reacts "without delay" (required for registration of an obstacle at the mobile robot.) The deliberative subcomponent must inform the reactive part of other possible solutions of a given situation.

The great part of research in the field of mobile robotics deals with hybrid agents design and applications, (see e.g. [2]), this architecture is mostly used for single robot systems.

1.2.4 "Behavior-Based" Agent

It is the most popular trend both in operation and robotics as well as design and applications of autonomous agents nowadays [32]. This agent employs a collection of concurrently executing behaviours decoding models of behaviour whereas each behaviour model represents goal-oriented block responsible for achieving certain task. Individual behaviour models typically consist of rules taking inputs from censors or other behaviour in the system and sending outputs to the effectors or other behavior models. Behaviour-based systems may be thus apprehended as structured networks of mutually interacted behaviours (*behavioural networks*). Behaviour inputs determine the behaviour activation status and are often defined as conditions that have to be executed in order to allow other behaviour to be used. In general we distinguish the following two types of activation conditions (behaviour preconditions):

- *Environment conditions*: Conditions that activate the behaviour based on a particular state of the environment.
- *Sequential conditions*: Usually defined as task-dependent and mostly represented by outputs of other behaviour models. Since it is possible to create so called *temporal sequences* (in this case they can be perceived as simple predefined strategies or plans) by means of behaviour activity chaining, the given behaviour is activated only when it is its turn in timely performing sequence. It is usually performed by mutual commitment among behaviour models (input of one behaviour is the output of the other behaviour).

Concept of behaviour-based architecture is the extension of reactive subsumption architecture. As mentioned before, reactive systems respond to current state of its local environment in real time by means of set of mutually concurring preprogrammed pairs "environment state → reaction". These reactive systems are rather limited since they lack the inner world representation which prevents them e.g. from keeping inner states plan or learning. One of the concept extension of the reactive systems which bring behaviour-based architectures is the ability to use inner representation whereas one of the significant contribution is the ability to use various forms of distributed representations instead of the central one. Both static structures and active procedural processes can be used as a individual forms of representations [7].

The use of abstract representations in a form in which they are utilized by deliberative architectures is the next extension adopted by behaviour-based systems. Abstract models have been established (abstract behaviours) that afford the opportunity to utilize individual primitive behaviours as manipulable operators in achieving partial tasks or goals [34].

1.3 Multi-agent System—MAS

Centralization or decentralization level is one of the most important properties at design and system modeling on the basis of multiple agents [25].

In general we may distinguish [8, 25, 38, 43]:

- Systems based on a single agent (centralized)
- Multiagent systems (decentralized, many decentralized systems utilize the role of leading agent, so called "leader")

Decentralized systems have many advantages over centralized systems, e.g. parallelism, fault tolerance, scalability etc. It is very difficult to find strictly centralized or decentralized systems in practice. Mostly we encounter hybrid systems where e.g. some central planner coordinates agents behaviour on the highest level of operation.

Single-agent systems seem to be more primitive than multi-agent systems. However the right opposite appears to be true. Distributed operation of multiple agents enables each agent to be modelled as primitive system whereas none of agents not have to be able to achieve the given task independently. There are no other entities recognized by an agent, thus if there were other agents in an environment, they would not be modeled as having their own goals, but as a part of environment. Multiagent systems differ from single agent systems in such a manner that each agent being in an environment is modeled as entity with certain goals, actions and skills. In general, direct mutual interactions of agents might appear in multi-agent environment. The other significant difference appears to be the fact that environment dynamics is directly affected by performance of agents. In addition, with regard to inaccuracy that may be typical for a particular domain, agents will act in an environment in an unpredictable way. Multi-agent systems may be thus considered to be dynamic.

Chapter 2
MAS Properties and Classification

The ability of agents to change their inner organization and to modify their interventions to environment is related to the following qualities: *adaptability and flexibility*. These abilities of agents depend on learning or mechanical (environmental) reasons that might affect the success of an agent at task solving.

The ability of system to dynamically adapt to environment changes and tasks is the core condition needed for creating a system which is *dynamically open*. This system should not consider to be the problem to change current state of agents engaged in the system within tasks solution process without giving any explicit note to other agents. At *static openness* of the system is the system able to change the number of agents in the task-solving process, however, all the other agents are informed of these changes. These systems, unlike dynamically open ones are not suitable for work in environment with high level of ambiguity. *Off-line systems* are referred when a task being solved come to a setback and the whole system is restarted at change of number of agents in the system. These systems are suitable for well-defined and well-predicted areas.

- Interference in the form of competition for shared means may appear in the systems composed of agents with the same goals. In the systems with agents of different goals, the interference may appear in the form of undesirable types of deadlocks and oscillations. In general, we can distinguish two basic types of interference from functional point of view [31]:
- Interference caused by multiplicity: *resource conflict, or resource competition*. It includes interference caused by competing of agents for shared sources and resources such as space, information, objects, etc. [8]. This type of interference increases with the growth of agents with the final effect of global behaviour quality decrease of a group. Resource conflict can be observed both in homogenous and heterogenous agent groups.
- Interference caused by *goal-related conflict or goal competition* is formed among agents with different goals. Agents might have comparable high-level goals, yet each individual agent can work with different potentially interacting subgoals.

M. Hajduk et al., *Cognitive Multi-agent Systems*, Studies in Systems, Decision and Control 138, https://doi.org/10.1007/978-3-319-93687-1_2

Homogeneity of agents is the attribute of agent group composed of agents with the same properties. *Heterogenous* system is, on the contrary, composed of agents with different properties. The level of heterogeneity of agents is closely related to the concept of task cover which enables agents to determine the level of ability to achieve task solution. This parameter is the evidence that cooperation is required and inevitable. In case the task cover is high the task may be solved almost (or completely) without cooperation. Otherwise, the cooperation is the key condition to perform the task. In the systems with homogenous agents, the cover of task is the highest and it decreases with the increasing level of heterogeneity. Communication, coordination and cooperation belong to the most important properties of multi-agent systems.

2.1 Agent Communication

The way communication is performed among individual agents in the system is one of the key elements at multi-agent systems structure design. Communication is one of *interaction* types of agents [8, 14]. Besides communication, *interaction by means of environment and interaction by means of sensors* is mentioned as well. In case there is no explicit type of cooperation, and cooperation among agents, the environment itself serves as the communication medium—the interaction through environment. It is the most primitive yet the most limited way of interaction. At censor-based interaction it is necessary for a robot to be able to recognize an agent (relative), and even after the censor-based interaction may be performed whereas there is no intention of explicit communication. The third type of interaction is communication which consists of explicit forms of intercommunication among agents by means of different message types.

Communication network (connects agents) and properties of individual agents determine *structure of multi-agent system* [30]. Providing the system has to solve given task class, it is required to meet three conditions:

- it should guarantee the *cover* in such a way that there is a group of agents (at east composed of one agent) for each key initial subtask which is able to solve it,
- agent should be able to pass its conclusions to the other agents while these agents are waiting for them and are prepared to utilize them later on depending on suitable connections,
- Sufficient computational resource (*potential*) and matching *communication paths* should be available in order to establish solution.

The key role of multi-agent unity is to make sure that *coordination, cooperation and communication* is performed among agents.

Since the architectures which support this form of communication use the same principles as standard communication networks, there were many technologies used in this field in order to create a design and formation of communication protocols. Their significant advantage is their efficiency, however, the system is unable to communicate with those agents who do not use the same communication protocol. This situation is easy to avoid by using more general communication protocol. In general, different types of multi-agent systems support Agent communication language such as e.g.

KQML: Knowledge Query and Manipulation Language ro *FIPA: Foundation for Intelligent Physical Agents.*

2.1.1 Communication Methods

The basic types of communication differ with reference to goal where the messages are lead within the unity of agents [30]:

- *direct communication*: messages can be sent directly to other agents (in synchronous and asynchronous mode)
- *indirect communication*: messages are concentrated in a preselected structure (buffer).

Methods of direct communication may be clearly characterized according to the number of potential receivers of initial message:

- *addressing mode of sending messages*: this method has many advantages regarding to utilization. When open systems are formed the number of agents in the system may dynamically change at the moment an agent disappears in the system and a message is being addressed.
- *omnidirectional message sending*: messages are sent to all the agents. Each agent in the system is replaceable. It is successfully utilized with adaptive systems and error-resistant systems. In case an agent dies it is replaced by the other. There are certain disadvantages, including the lack of information security —each agent has access to any message. In the systems with higher intensity of sending messages, communication infrastructure may be easily overloaded as well as excessive loading of each agent might occur. From the programming point of view, the non-directive message sending is complex and inconvenient.
- *selective message sending*: each agent is the member of at least one group of agents. This kind of system divided into groups has the communication structure composed of addressing mode of sending messages to groups. All agents which belong to a target group of agents have access to these messages. This method of message sending focuses on combination of previous two methods.

2.2 Coordination and Cooperation

The main goal is to distribute tasks among agents in an appropriate way. Global behaviour of multi-agent system is highly affected by the ability of agents to mutually coordinate their individual actions by means of which they maintain common goals. Cooperation is often considered to be one of the key attributes which distinguishes multi-agent systems from relative disciplines such as distributed computing, object-oriented systems etc. In spite of the fact that agent systems cooperation is often used, the meaning of this notion is ambiguous. Cooperation methods concentrate on designs of organization principles referring to smaller agent group selected at carrying out certain task which insure its successful completion [30]. Coordination and cooperation might be viewed as two sides of the coin—the decision of one of these areas compels the utilization of certain procedures in the second one and vice versa.

There have been [42, 43] defined four generic phases of cooperation problem-solving process based on different systems generalisation for cooperative task-solving:

- *identification*: The process itself starts with identifying the need of cooperative action. The identification may be carried out by e.g. the determination of an agent that its abilities are insufficient for achievement of currently solved goal independently, i.e. agent needs some form of assistance for maintaining its goal.
- *team formation*: An agent which has found out that cooperative action is required is trying to gain the assistance of the other agents. In case the phase is successful, the agent group is ready to solve the problem task collectively.
- *plan formation*: Agents are trying to assemble an action plan (in an negotiative way) which would help them reach the goal while the plan has to represent the compromise of their intentions.
- *team action*: The current version of setting up common action plan is being carried out. There are certain relations maintained among agents at plan performing process that are defined by fixed social conventions the agents observe.

2.3 Mobile Agents

A mobile agent is not linked to a system where the compilation process of its code begins. It is capable of migration from one system in network to another. Its ability of migration enables it to move to a system which contains an object it wants to communicate with and thus utilize the advantages of location in the same computer or the same network as a given object [25].

It might seem that this notion is closely related to mobile robotics, but the mobile agent is not an agent on physical (hardware) level. In fact, it is a program (software) capable of transmission of its source code through network from one computer to another. The agent's location is the environment where performance of its code is carried out. The location is assigned by a computer with IP address where the mobile agent is located at the time of performance, by operating system the given computer works under and by mobile agents message system which provides the mobile agent with its own compilatory environment with further required functions. General model of mobile agent is depicted in Fig. 2.1.

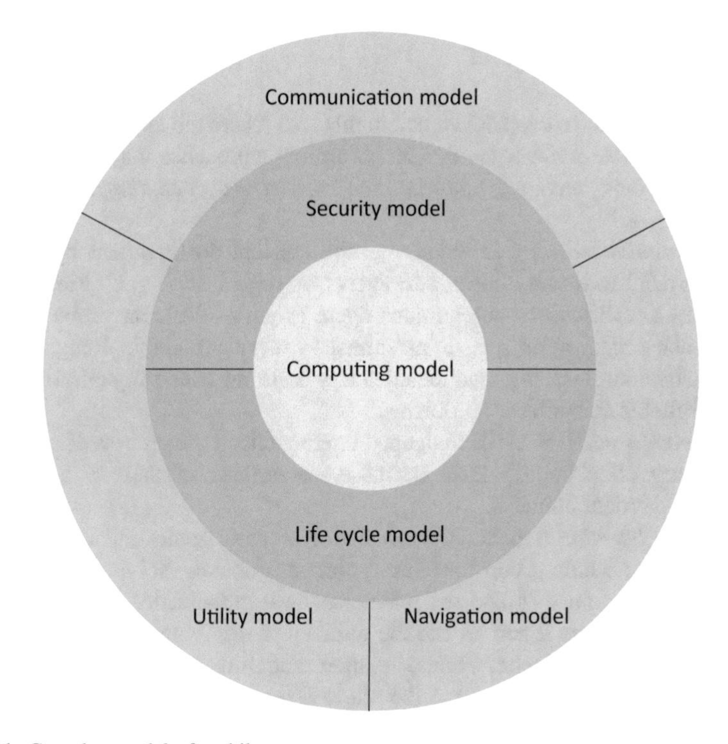

Fig. 2.1 Complex model of mobile agent

Computational model: the operation of its procedure is carried out by physical processor or virtual processor.

Utility model: an agent is perceived from the point of view of functions it has to perform for a guide it belongs to. It is a general agent model, where mobility is only a complement.

Lifecycle model: represents a mobile agent's states it can get into during its lifetime.

Navigation model: characterized by the process from an agent's request to move, the transport and adoption of an agent at a given place together with regeneration of its running.

Communication model: represents the need of sending messages by means of orders.

Security model: summary of strategies to minimize the imminent danger from both other agents and mobile agent message system.

2.4 MAS and Learning

Learning methods have been developed in this area where the learning subject is not an agent, but whole multi-agent system. Learning approaches may be divided into five categories: *organisation, coordination, coherence, communication, and distributed learning*.

Organisational learning is defined as the concept development by means of which the social roles are allocated to agents by organisation [11]. Single organisation types are formed in autonomous agent group at different problem-solving whereas each organisation type is presumed to make problem-solving more efficient. Organization learning thus includes also learning which organisation type is the most suitable for different problems.

Coordination skills are skills of agents' interactions in the process of mutual task solving. They often contribute to decrease the number of these interactions by excluding redundant contacts.

Coherence learning is behaviour adaptation leads to achievement of a suitable behaviour of the whole group for current mutual problem—solving.

System is able to use "broadcast" of messages at initialization process, however even after certain time it can reduce the number of receiving bodies of individual messages by learning of other agents, i.e. their modelling to such a state that all the messages will only be addressed to the relevant receiving bodies—*communication learning*.

Distributed learning: Agents can distribute their skills among other agents. Agents can thus learn new things.

2.5 Hierarchical Architecture of MAS

Multi-agent architectures (structures) may be divided due to several criteria, e.g. time period the interactive relations are set, present agents architecture, used communication, etc. The following categorization will be focused on tasks division and further centralization of data and operation.

2.5.1 Centralized Multi-agent Systems

There is one superior agent which communicates with all subordinate agents. Subordinate agents do not communicate among each other. The central agent cannot be added to the system and addition of subordinate agent requires reprogramming of the central agent. Complexity of subordinate agents is smaller compared to the central agent that disposes of all the information. Example of centralized agents arrangement may be seen in Fig. 2.2.

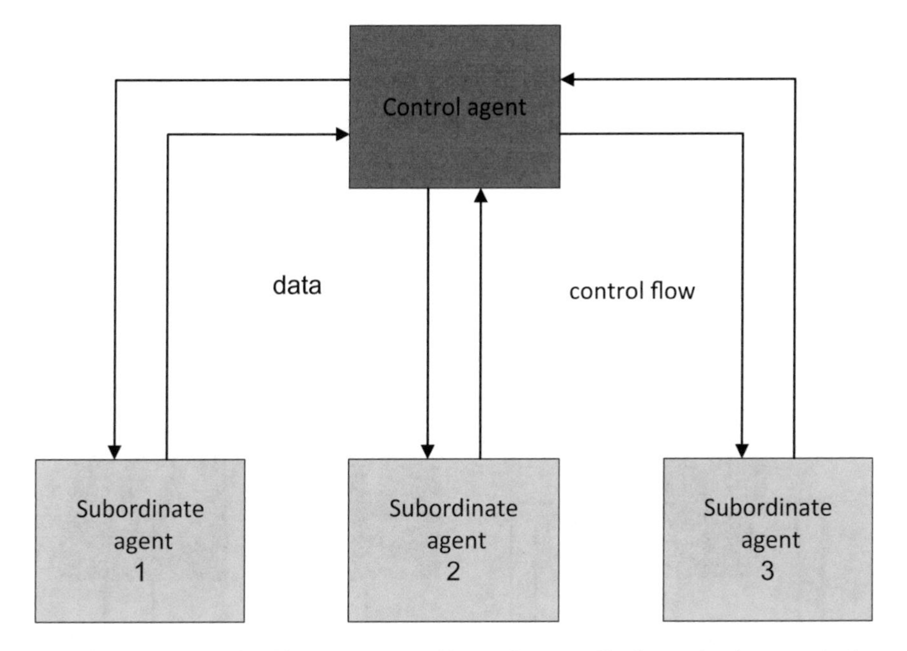

Fig. 2.2 Architecture of multi-agent systems with complete centralized control and communication

2.5.2 Hierarchical Multi-agent Systems

Alike the previous architecture, there also exists central authority. However, operation and communication are communicated to lower organization levels. Data and control information are communicated only in vertical direction. Horizontal communication is not possible. In case of dropout of one of the agents, the impact on functionality depends on position of an agent in organization hierarchy. Adding an agent into system requires configuration change of its superior agent. Communication flow of data and control is rather evenly distributed in all multi-agent system and it carries the increasing function of organization degree. Example of the hierarchical arrangement of agents in MAS may be seen in Fig. 2.3.

2.5.3 Federated Multi-agent Systems

There does not exist control centralization, just task centralization. Drop-out of a mediating agent represents a threat to whole application. It is required that its adding into system is communicated to user agents. Communication demands are high since the communication among agents is in progress. Mediating agents decrease the charge. Example of federated arrangement of agents in MAS may be seen in the Fig. 2.4.

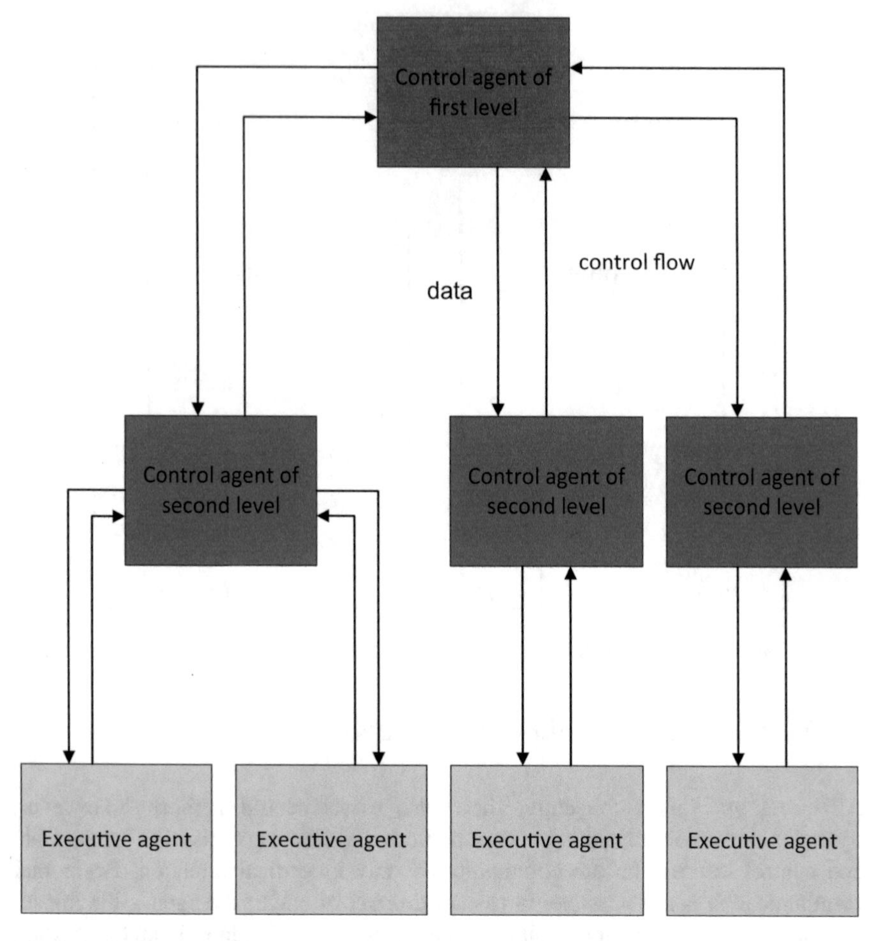

Fig. 2.3 Hierarchic architecture of multi-agent system. Several agents of different hierarchy levels can present its organizing part

2.5.4 Decentralized Multi-agent Systems

This architecture is also considered to be autonomous. Its basic characteristic feature is task decentralization and control in the system without any evidence of mediating or central element. The level of influence on system features concerning static or dynamic relations in the structure is high. Dynamic decentralized systems

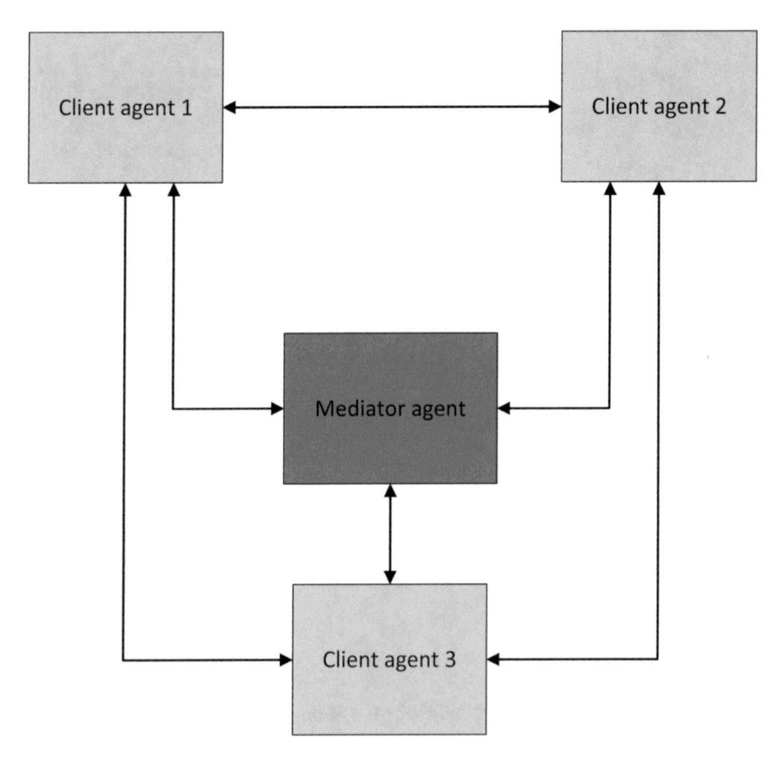

Fig. 2.4 Architecture of federative (unite) multi-agent system

finds it much less serious to lose an agent. Its negative feature is complex coordination of activities, It is immensely difficult to find a solution to the complex coordination of activities comparing to hierarchical arrangement of agents. The example of decentralized arrangement of agents in MAS may be seen in Fig. 2.5.

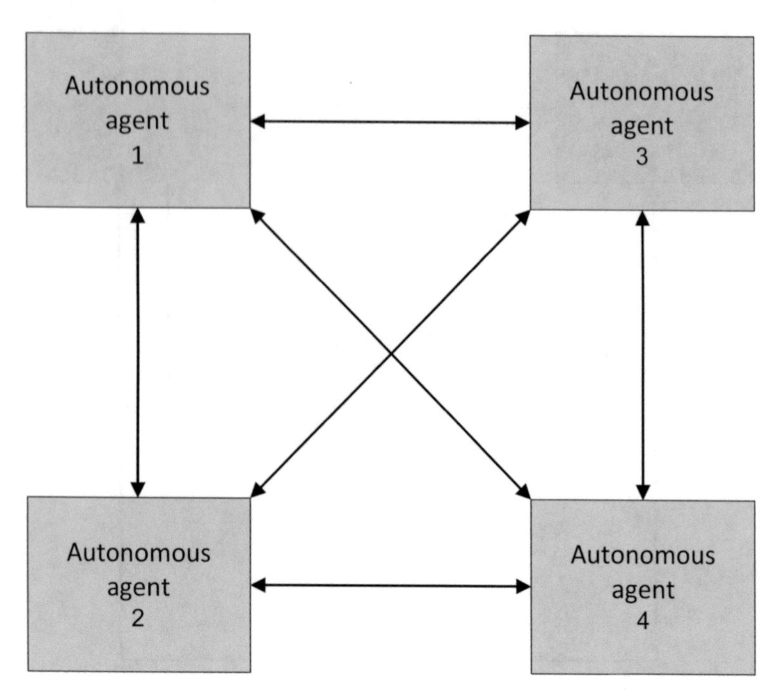

Fig. 2.5 Architecture of decentralized multi-agent system

Chapter 3
Agent Approach to Multi-agent Systems

Multi-robot systems are becoming more and more popular and useful in the field of industrial production, lifesaving operations, the Earth and the Space research military operations, agriculture, entertainment industry, etc. Application of more robots for task solving from these areas bring many advantages comparing to solutions which depend on one robot only [77]:

- **parallelism**—employment of various robots shortens the overall time of task solving process. Space mapping and browsing or editing may serve as a good example.
- **robustness**—in a case when one of the robots became unable to attend certain task (failure) its task may be taken over by the other robot with similar functionality and the whole system keeps achieving goals. In some applications, especially military ones, it allows for destruction of some robots therefore it is required to employ more robots that can be substituted.
- **price**—costs required for the production of one complex robot may exceed the costs needed for the production of more simple robots. In case of more complex robots there is a problem with reconfiguration since they are designed for a specific task. Simpler robots may be reused as a part of team in various applications.
- **scalability**—this factor is closely related to the price. In case the parameters are changed (e.g. browser surface extension) is often easier to extent the team by adding one or more robots than to adjust properties of one robot in order to meet the requirements to task solving process.

Some tasks in their fundamentals cannot be accomplished by one robot. These are tasks which require physical presence of robots on different places in the environment at the same time, such as long distance information transmission in real time whereas robots must create communication bridge. On the other hand, the advantages of using multi-robot system are well-balanced by higher complexity at the process of design, implementation and testing of such system.

M. Hajduk et al., *Cognitive Multi-agent Systems*, Studies in Systems, Decision and Control 138, https://doi.org/10.1007/978-3-319-93687-1_3

3.1 MAS and Multi-robot Systems Relations

Methods and principles from the field of multi-agent systems are often used to design organization structures and controlling algorithms of multi-robot systems [77]. The algorithms originally developed for highly-distributed agent decision-making are often used for robot group control. These are mainly distributed action planning algorithms, negotiation mechanisms, and communication languages and protocols required for communication among agents.

Using multi-agent technologies in the field of robotics brings also problems and restrictions which we have to bear in mind at the moment of system architecture and control algorithm design. These restrictions are the consequences of the fact that robots are physically existing entities and work in physically existing environment [77]:

- During the process of planning and action operation it is necessary to consider communication and location of robot in space while other information agents are not usually limited with who they will communicate. Communication range is often limited in robotic applications and so is also the selection of potential communication partners.
- robots operate in real time—in some cases it is necessary to prefer reaction speed on the given situation to the complexity of problem consideration and looking for optimal solution. Sometimes it is advisable to limit communication intensity because relatively slower actions of a robot in the physical environment.
- agents usually work with skills represented in symbolic form while robots obtain the environment information by means of their censors. It is necessary to process this information at first, and change it into symbolic form after.
- the real world is dynamically changing and these changes cannot be predicted. This is the reason why we have to emphasize robustness of algorithm in case of failure of some source or action when unexpected obstacle is formed, etc.
- robots must perform a lot of supporting activities together with achievement of their goals such as localization, mapping, production of environment models. All these activities are memory and computing challenging. This fact must be considered at the optimization of communication range and algorithm referring to negotiation among robots.

Chapter 4
Multi-agent System Test Domain—Robosoccer

In 1990's two robotic competitions were created independently. In both competitions robots compete against each other in disciplines by imitating soccer game. They were formed because of possibilities of testing designed and applied solutions within robotics and artificial intelligence in general. Scientific teams from all over the world might test the correctness of their solutions applied on a group of mobile robots playing soccer in direct confrontation within entertaining game. There are international competitions in both games taking place every year which bring robotics to spectators in attractive and entertaining way. The organizations covering these competitions are called *RoboCup* and *FIRA*.

The final goal of *RoboCup* initiation is to develop fully autonomous humanoid robot team that will be able to beat the world champion team in traditional soccer in regular soccer match by 2050 [72].

The idea of robots playing soccer was mentioned in 1992 by professor Alan Macworth (University of British Colombia, Canada) for the first time in the article "On Seeing Robots" issued in the book Computer Vision: System, Theory, and Application. Independently, the group of Japanese scientists organized seminar about great challenge at artificial intelligence in October 1992 in Tokyo. The scientists seriously discussed the use of soccer as a game for science and technology advertisement. Subsequently, the feasibility study and costingness study were performed. After the conclusion was made that the project can be realized and is desirable, the rules and prototypes of soccer robots and soccer simulation were developed. In June 1996 the group of scientists including Minora Asada, Yasuo Kuniyoshi and Hiroaki Kitano organized the robotic competition named Robot J-League. After a short time, with regard to its popularity from all over the world, the project was renamed to Robot World Cup Initiative (Robocup). In November 1996 the Pre-RobotCup-96 took place in Osaka (Japan) where 8 teams competed in simulation league. The middle size league was demonstrated as well. The first official RoboCup competition took place the following year (August 1997, Nagoya—Japan) which happened to be extremely successful. More than 40 teams participated and more than 5000 spectators watched it [72] (Fig. 4.1).

M. Hajduk et al., *Cognitive Multi-agent Systems*, Studies in Systems, Decision and Control 138, https://doi.org/10.1007/978-3-319-93687-1_4

Fig. 4.1 A humanoid playing football at RoboCup project [62]

FIRA (Federation of International Robot-soccer Association) was established in June 1996. Even sooner in 1995, the professor Jing-Hwan Kim (Korea—Advanced Institute of Science and Technology KAIST, Korea) organized the committee meeting for Micro-Robot world Cup Soccer Tournament (MiroSot). So called pre-meeting in category MiroSot was held throughout 29th June–4th August 1996 where 30 teams from 13 countries participated. The first MiroSot'96 tournament was held between 9th November–12th November 1996 (KAIST) and 23 teams from 10 countries attended this tournament. As early as in June 1998 FIRA Cup (Paris, France) was held where another categories were joined. Nowadays FIRA organizes competitions in the following categories [67]:

- HuroSot (Humanoid Robot World Cup Soccer Tournament)
- AmireSot
- MiroSot
- NaroSot—menšia verzia MiroSotu
- AndroSot
- RoboSot
- SimuroSot.

Slovakia won the first prize in 2010 (Bangalore, India) both in 5 versus 5 and 11 versus 11 competition (number of robots 5 or 11) (Fig. 4.2).

Fig. 4.2 MiroSot, particularly 11 versus 11

4.1 MiroSot Category

The most dynamic matches between teams in real time may be seen in this category. The robot speed might sometimes exceed 3 m/s and their acceleration might reach about 10 m/s^2.

The image processing is provided by a camera. The image is transmitted to control computer where it is processed. Strategy is then generated from processed image (teams mostly implement MAS). Finally, strategic instructions to robots—players are sent by means of high frequency communication. Each team has to dispose of information transmission carrier frequency that can be modified (at least two different frequencies) to avoid the situation that teams competing with each other would send data with the same carrier frequency and they would disturb each other. The size of each robot is limited by cube of edge 75 mm. The only exception is the aerial that is out of size. The basic architecture of used hardware in the MiroSot category may be seen in the Fig. 4.3.

There exist two types of league in this category nowadays: the small size robot soccer and the adult size robot soccer leagues. The Small Size robot soccer game takes place between two teams of five robots each on the field that fits the size 220 cm × 180 cm. The size of the adult size robot league field has to fit within the size 420 cm × 280 cm. the behaviour realization of the whole system (soccer team), hardware configuration, the way of communication and mobile robots solution are not limited, however, each teams chooses the best alternative for them (experience of realization team, expenses at team building, etc.). The complete description, rules and restrictions can be found at organisation FIRA home page [67].

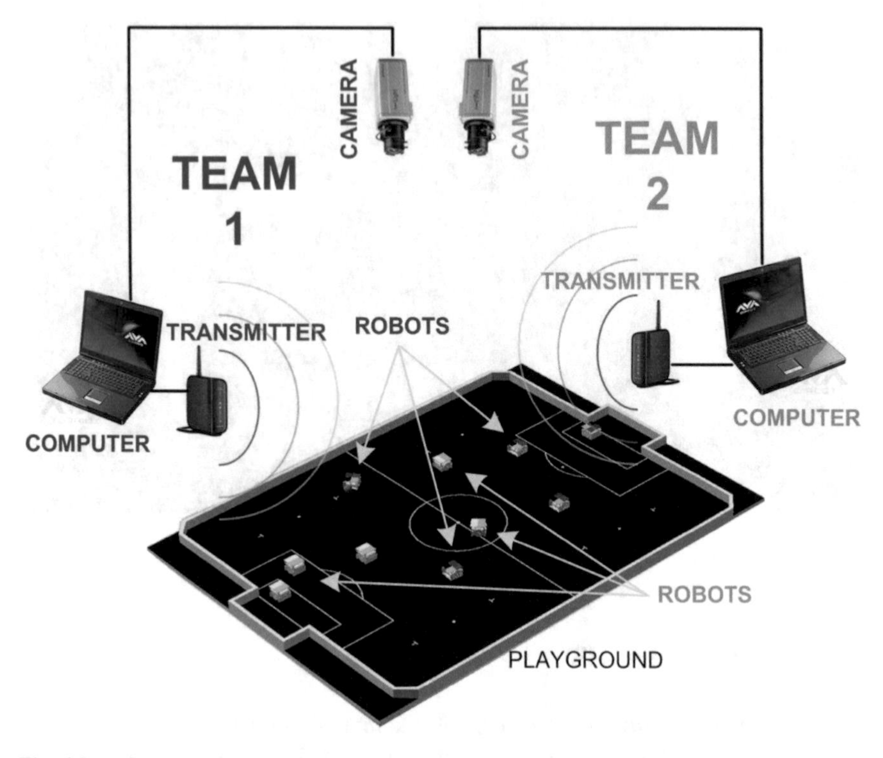

Fig. 4.3 Robosoccer architecture

4.2 The Function Block Application Process and Active Interventions in the System

The control cycle performed by different teams in category MiroSot may be generalized and described in the following repeated sequence (Fig. 4.4):

- sampled scene (playground)
- image processing
- strategic planning (computing)
- communication between control computer and robots
- movements of robots following received instructions
- scene manipulation by the movement of robots.

4.3 Mirosot and MAS

Most of the teams use multi-agent approach to compute strategies in this category. The organizational structure, communication, cooperation, of such MAS is dependent on basic philosophy of system arrangement. MAS attributes depends on

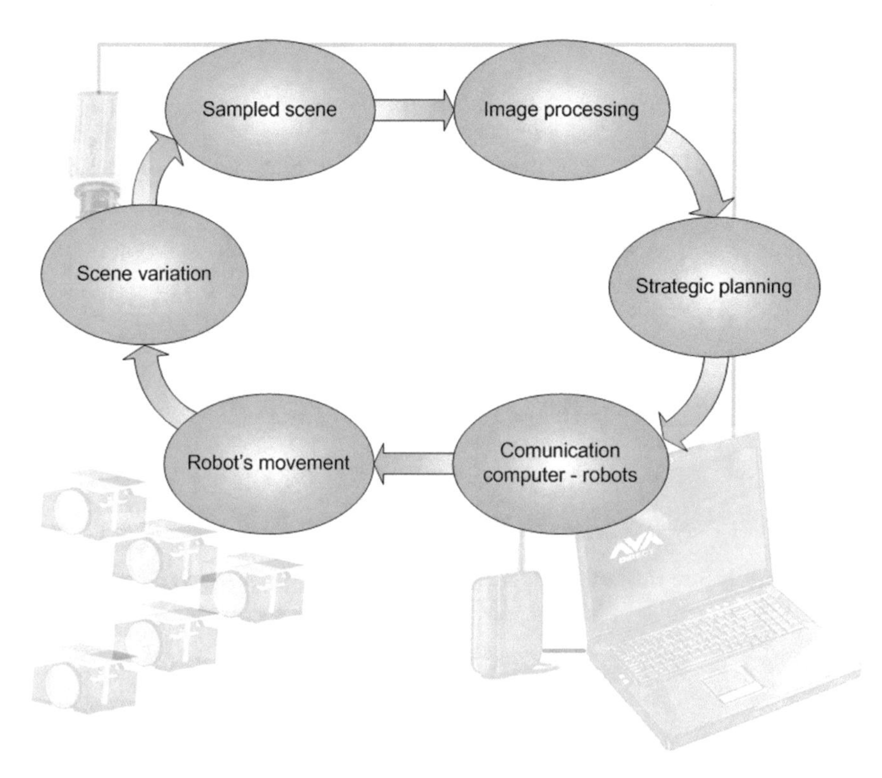

Fig. 4.4 Action sequence in control loop in Mirosot category

type and complexity of applied agents. The most popular is the structure with one superior agent (coach or captain, Figs. 4.5 and 4.6) and decentralized structure of hierarchical organization of agents.

Decentralized structure (Fig. 4.5) is famous for having all the agents on one hierarchical level. None of the agents is superior or inferior. The inner structure of agents is complex. These agents are mostly hybrid or model-based reflex.

Centralized structure (Fig. 4.6) allows for simplification of agent players and delegates some tasks to the central agent. The arrangement of agents may have the

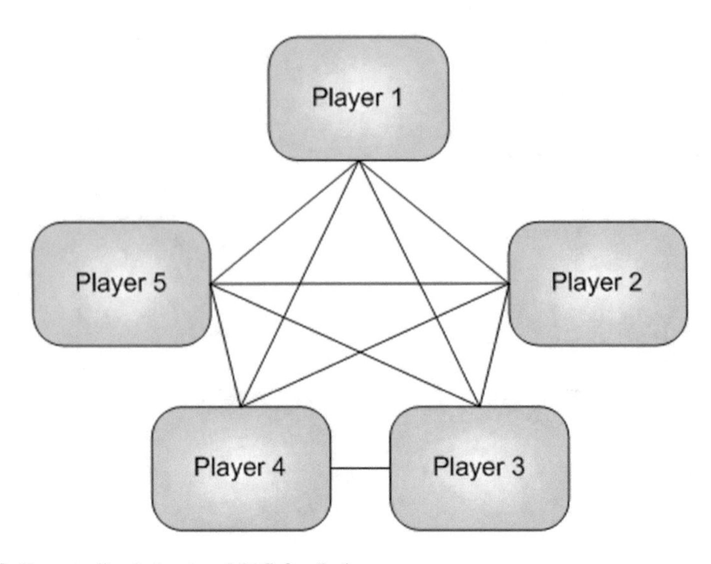

Fig. 4.5 Decentralized structure MAS for 5 players

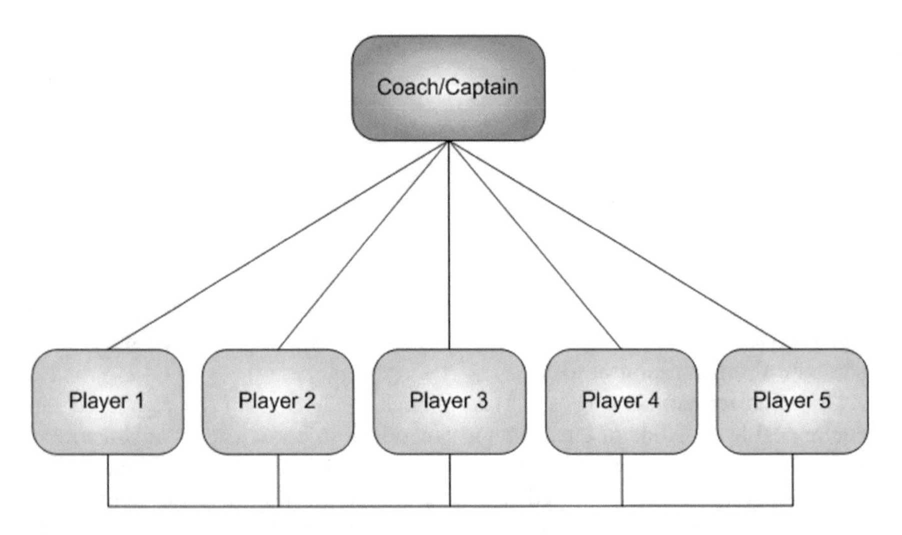

Fig. 4.6 Centralized structure MAS with one superior agent

following structure: the central agent solves strategy, i.e. it is model-based reflex and the players are reactive, or hybrid which involves track planning.

The individual teams employ hierarchical arrangement similar to these two structures or their modifications.

Chapter 5
Multi-agent System Application in Robosoccer

SjF TUKE Robotics team which attends both European and International competitions organized by the international organization FIRA (Federation of International Robot-soccer association) has described the approach towards the problem-solving process of robot cooperating group in this chapter. Multi-agent approach was used for mobile robots (players) group control solution. Each **agent —robot** is a merge of the elementary **agent—player** (agent's mind—software concept of the agent) and **robot—player** (body—construction, electronic hardware and software) (Fig. 5.1). There have been created 12 elementary agents—players throughout the team development so far. Each elementary agent has designated subtask on the field which helps them achieve multi-agent system goal.

Complex image of applied software and hardware modules and their relation to agents in designed system is displayed in the Fig. 5.2 where blocks with round edges represent hardware modules, and blocks with sharp edges represent software modules. There are also relations among individual blocks presented in the module arrangement design. And thus the control loop process is given (downwards).

Throughout the above mentioned general control cycle of each team, there is a camera at the beginning of module chain. The image is transmitted to image processing software module by means of interface IEEE 1394a (firewire). The result of image processing is represented by the positional parameters of both robots and the ball. All these parameters are saved in memory (shift buffer) of size for 67 last frames. This memory is then accessible to the agent (master) and the elementary agents (players). The agent (master) uses this memory for estimation of the movements of the ball and the opponent and is able to select from the defined strategic actions. 5 elementary agent players are chosen by the selected strategic action. These players were allocated to the given actions during the phase of development within the overall strategy. Finally, the robot player (ID) is given to the elementary agents. The agent master decides on the selection of the strategic action and the selection of the elementary agent players pursuant to the underlying strategy chosen by user. The elementary players dispose of all the information

M. Hajduk et al., *Cognitive Multi-agent Systems*, Studies in Systems, Decision and Control 138, https://doi.org/10.1007/978-3-319-93687-1_5

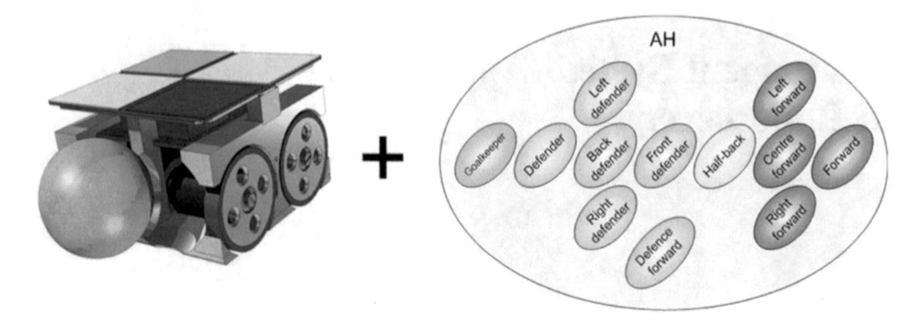

Fig. 5.1 Formation of complex robot-player joining together with robot-player and one of elementary agents

required regarding to the ball and all the robots location on the playground. They plan their movement based on the information in order to get the ball into the opponent's goal or to prevent the opponent from scoring a goal. After planning their movement they send the information to software module which provides the communication. After data have been adjusted in order to be joined to robot players, they are then transferred to hardware module via USB for interface transformation to RS485. Each robot uses the transmitter with its own set number of used channel to transmit data in a wireless manner. Data are then cyclically sent to each robot player in parallel, until new data are recorded. Robot player changes its speed parameters required as soon as possible after transferred data package has been recorded. This means the scene change for the next recorded frame.

5.1 The Camera

One of the most significant requirements at selecting the camera was its frame frequency, high-quality and stable frame production. From all the suitable cameras, AVT camera Marlin F-033C has been used (Fig. 5.3).

The frame rate (frames per second) of 568 $\times$ 426 size is 66.7 fps which means that a new frame has been captured every 15 s. Within this time, at least the image is to be processed to allow strategy for reconfiguration.

5.2 Image Processing

In order to decrease the delay from the moment of loading a frame to action realization by the robots there was used simple algorithm of image processing. Image processing follows saving of captured and transferred frames to memory of the controlling computer. ActiveX component (DirectFire package) provided by the

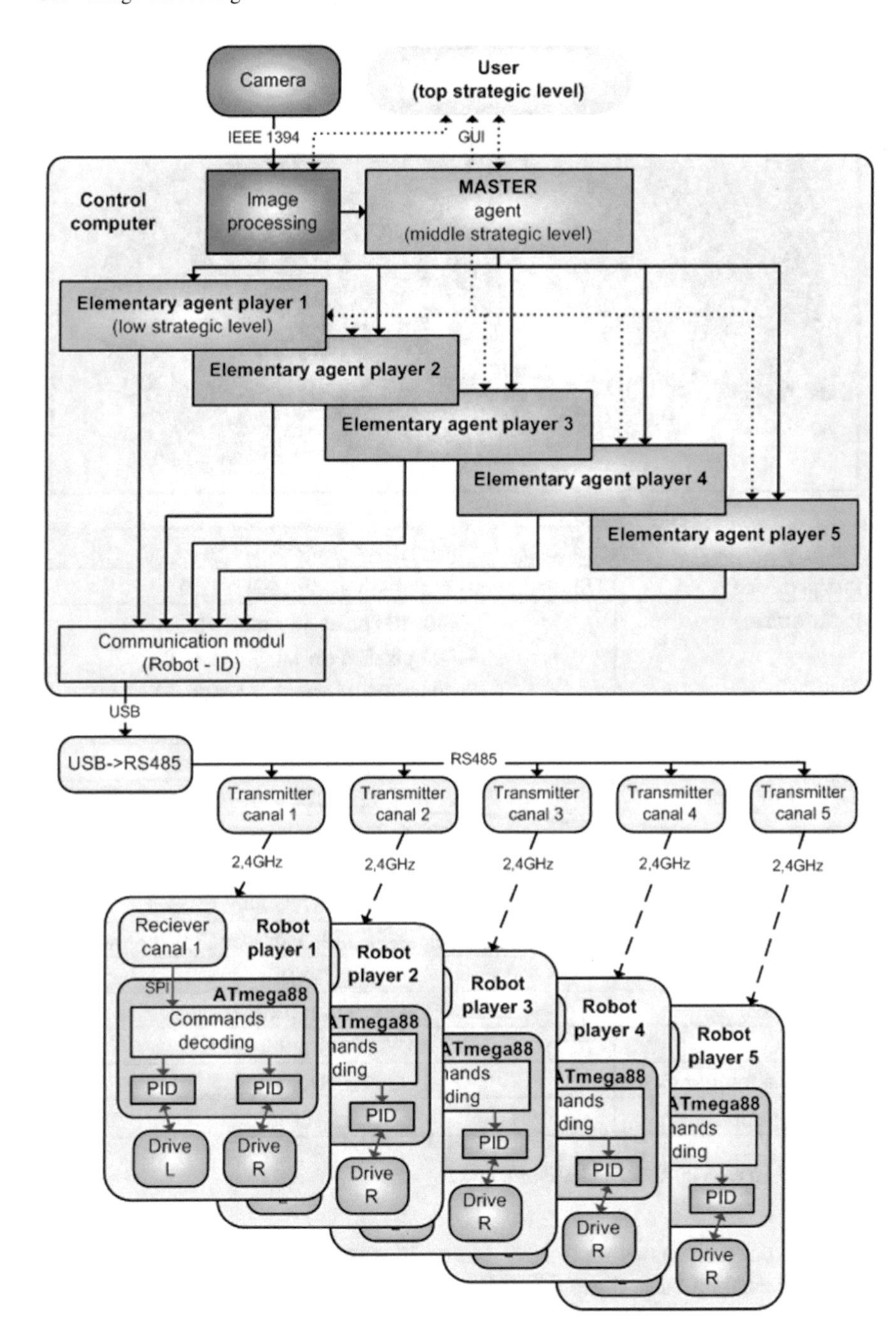

Fig. 5.2 Complex visualisation of used hardware and software modules

Basic parameters of the used camera	
Image device	1/2(diag.8mm) progressive scan, SONY CCD
Picture size	640x480 pixels (Format_0) 656x492 pixels (Format_7 Mode_0) 656x494 pixels (Format_7 Mode_1)
Cell cize	9.9um x 9.9um
Resolution depth	8bit/10bit (b/w only); 12bit (ADC)
Digital interface	IEEE 1394 IIDC v. 1.3
Transfer rate	100Mbit/s, 200Mbit/s, 400Mbit/s
Frame rates	68Hz (YUV 4:1:1); up to 50Hz (YUV 4:2:2); 33Hz (RGB)
Shutter speed	20...67.108.864us (67s); auto shutter
Power requirements	DC 8V-36V through IEEE 1394 cable or 12pin connector
Power consumption	< 3W (12VDC)
Dimensions	72x44x29 mm (LxWxH)
Mass	< 120g (without lens)
Operating temperature	+5 .... +45°C
Storage temperature	-10 .... +60°C

Fig. 5.3 Used camera AVT Marlin F-033C [61]

camera producer enables us to save frames in memory in three two-dimensional fields representing basic colours B [568], [426] … red, G [568], [426] … green, B [568], [426] … blue). DirectFire package provides indexes to the beginning of these fields. When the camera is set in such a way that edges of the field are on the boundaries of camera's field of view, it means 4.5 mm/point. At robot's edge length of 75 mm the average number of pixels approximately equals to 16. After rotation

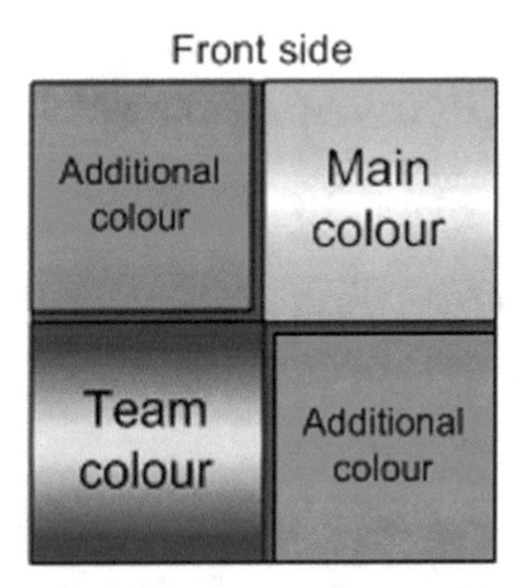

Fig. 5.4 The pattern recognition of the robot players

of the robot at 45 degree, its diagonal length is 106 mm, which equals to 24 pixels. In order to comply with the rules, the selected pattern of identifying robots is displayed in the Fig. 5.4.

Complementary colour may be on two possible positions. It is possible to create even eight patterns as it is displayed in the Figs. 5.5 and 5.6 by combination of colours for one team colour.

Fig. 5.5 Colour combination for yellow team colour

Fig. 5.6 Colour combination for blue team colour

The proposed and used algorithm of processing is not appropriate for the industry use where high reliability of photogrammetrical plotting is required. Algorithm detects the positions of the objects from the original frame of the field (Fig. 5.7) It was composed of three basic parts:

- detection of used colours (Fig. 5.8).
- detection of colour clustering
- detection of objects (Fig. 5.9).

Fig. 5.7 The basic image achieved by ActiveX DirectFire package component

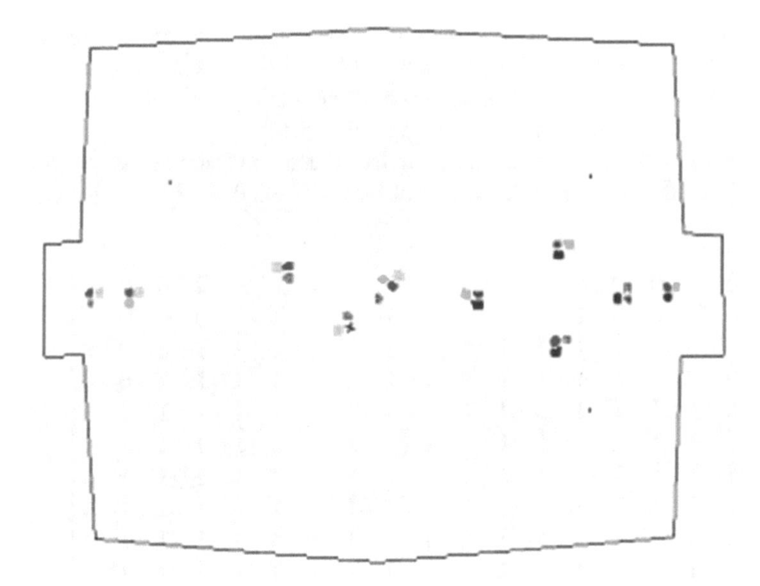

Fig. 5.8 Used colour detection

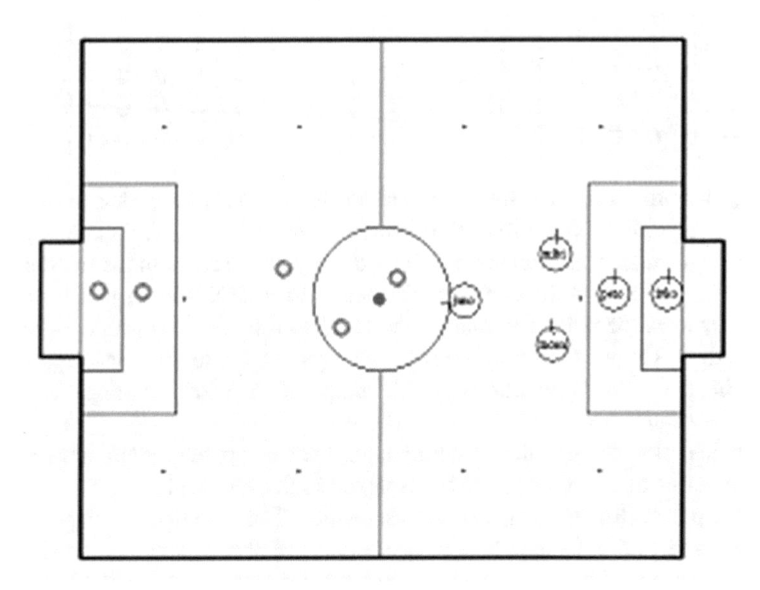

Fig. 5.9 Object detection

Colour detection is understood as the selection of colours that are used for distinguishing the robots from the ball. The selection takes place at embedded cycles where the colour of each pixel is compared to R, G, B colour models which were set up by the user in colour settings (Fig. 5.10).

After the colour selection, the module for colour clustering detection comes. It was created by experimental setting of browsing window given by the clustering matrix.

$$\begin{vmatrix}
0&0&0&0&0&0&1&1&1&1&1&1&0&0&0&0&0&0\\
0&0&0&0&1&1&1&1&1&1&1&1&1&1&0&0&0&0\\
0&0&0&1&1&1&1&1&1&1&1&1&1&1&1&0&0&0\\
0&0&1&1&1&1&1&1&1&1&1&1&1&1&1&1&0&0\\
0&1&1&1&1&1&1&1&1&1&1&1&1&1&1&1&1&0\\
0&1&1&1&1&1&1&1&1&1&1&1&1&1&1&1&1&0\\
1&1&1&1&1&1&1&1&1&1&1&1&1&1&1&1&1&1\\
1&1&1&1&1&1&1&1&1&1&1&1&1&1&1&1&1&1\\
1&1&1&1&1&1&1&1&1&1&1&1&1&1&1&1&1&1\\
1&1&1&1&1&1&1&1&1&1&1&1&1&1&1&1&1&1\\
1&1&1&1&1&1&1&1&1&1&1&1&1&1&1&1&1&1\\
1&1&1&1&1&1&1&1&1&1&1&1&1&1&1&1&1&1\\
0&1&1&1&1&1&1&1&1&1&1&1&1&1&1&1&1&0\\
0&1&1&1&1&1&1&1&1&1&1&1&1&1&1&1&1&0\\
0&0&1&1&1&1&1&1&1&1&1&1&1&1&1&1&0&0\\
0&0&0&1&1&1&1&1&1&1&1&1&1&1&1&0&0&0\\
0&0&0&0&1&1&1&1&1&1&1&1&1&1&0&0&0&0\\
0&0&0&0&0&0&1&1&1&1&1&1&0&0&0&0&0&0
\end{vmatrix} \tag{5.1}$$

The clustering matrix is always translated with 2 steps jump due to the acceleration of algorithm. The size of the clustering matrix (18×18) was chosen as an arithmetic average of the number of pixels displaying the robot in the middle of the field (20×20) and in the corner of the field. (16×16). With regard to fish eye caused by lens, the objects located in the middle of the field take more space than on its edges. Colour clustering search is always subject to the cluster pixel (if a pixel in the clustering matrix = 0 → jump, if a pixel in time clustering matrix → actual_ colour ++).

The final view through the clustering matrix and computing of colour centres for different cases of robot's rotation is displayed in the Fig. 5.11.

Black points display computed colour centres. The arithmetic average of coordinates x and y of all the points of the same colour is the colour centre in clustering matrix. In case all three colours located in the matrix are in the amount over the set threshold, then the robot has been detected. If more robots with the same colour combination have been detected, then the pattern with higher sum of pixels of team and main colour is preferred. The calculation of position and rotation is defined by the robot's centre and the angle of rotation φ Fig. 5.12).

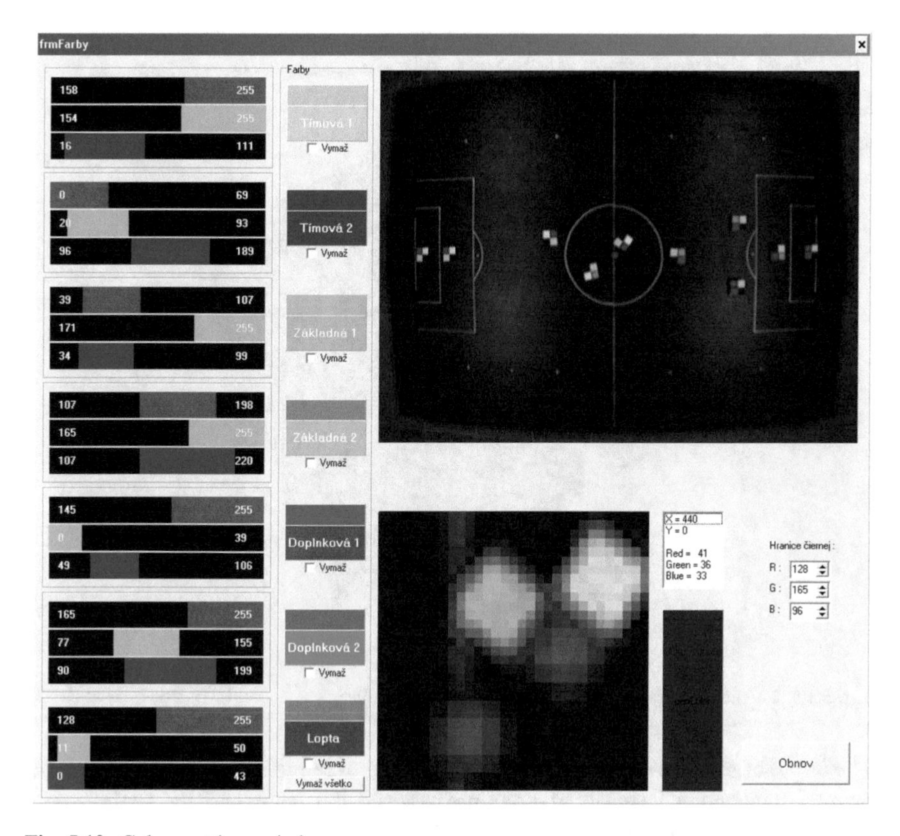

Fig. 5.10 Colour settings window

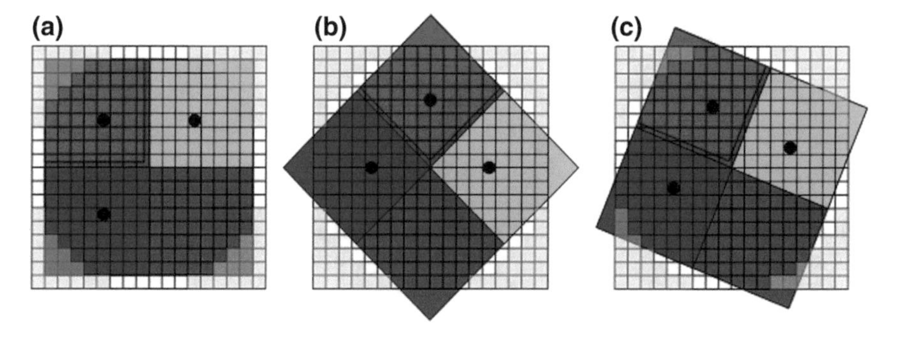

Fig. 5.11 A view to a robot player with different orientation (a, b, c) via cluster matrix

The physical centre of the robot is defined by the centre of flowline centroids of team and main colour. The vector product of two vectors leading from the centre towards the centroid of main or complementory colour defines the position of the complementary colour. Thus the specific pattern of represented robot is defined.

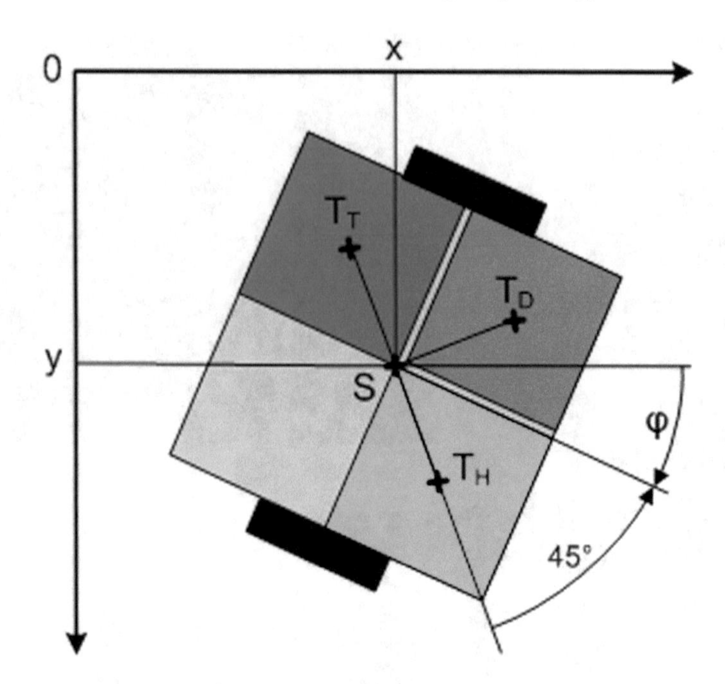

Fig. 5.12 Physical centre and rotation angle of robot player

The angle of rotation of robot player φ compared to reference coordinate system is defined by angular rotation of flowline between the team and the main colour around the centre in 45°. In the end all the robot coordinates are transformed in accordance with parameters which are set by the user during the process of camera adjustment:

Pinch: fish eye removal (K)

$$x' = \frac{x}{1 + K \cdot r} \qquad y' = \frac{y}{1 + K \cdot r} \tag{5.2}$$

Zoom: enlargement or reduction of an image (Zoom)

$$x' = Zoom \cdot x \qquad y' = Zoom \cdot y \tag{5.3}$$

Angle: the image rotation around the frames' centre (angle)

$$x' = x \cdot \cos(Uhol) - y \cdot \sin(Uhol) \qquad y' = x \cdot \sin(Uhol) + y \cdot \cos(Uhol) \tag{5.4}$$

Shearing (PX, PY)

$$x' = x + (PX \cdot x \cdot y)$$
$$y' = y + (PY \cdot x \cdot y) \tag{5.5}$$

Shifting (Shift X, Shift Y)

$$x' = x + Posun\,X$$
$$y' = y + Posun\,Y \tag{5.6}$$

It is necessary to transpose the pixels (Eqs. 5.2–5.6) to make sure that the image centre has zero coordinates before the pixels have been transformed.

$$x' = x - \frac{Pixel\,Width}{2}$$
$$y' = y - \frac{Pixel\,High}{2} \tag{5.7}$$

It is necessary to perform the feedback with these Eqs. (5.2–5.6) after the transformation:

$$x' = x + \frac{Pixel\,Width}{2}$$
$$y' = y + \frac{Pixel\,High}{2} \tag{5.8}$$

Defects brought by lens, indefinite positioning of the camera over the centre of the field and angle discrepancy of the camera from vertical axis. Figure 5.13 shows the field frame before and after the image transformation.

5.3 The Agent MASTER

It is software module. Its key function is to transform the basic strategy chosen by the user to the strategic actions. It uses created rules defined for each strategic action allocated to the basic strategy (Chapter 1). Following the rules, the master would utilize exactly five elementary players (totally there were 12 elementary players formed) which are able to achieve goals for the specific strategic action.

The master is in fact the type of software agent which perceives the field through the module of image processing and is able to affect the future state on the field via elementary agents. By affecting the positions and tasks of robot players, it is also capable to dynamically affect the game development.

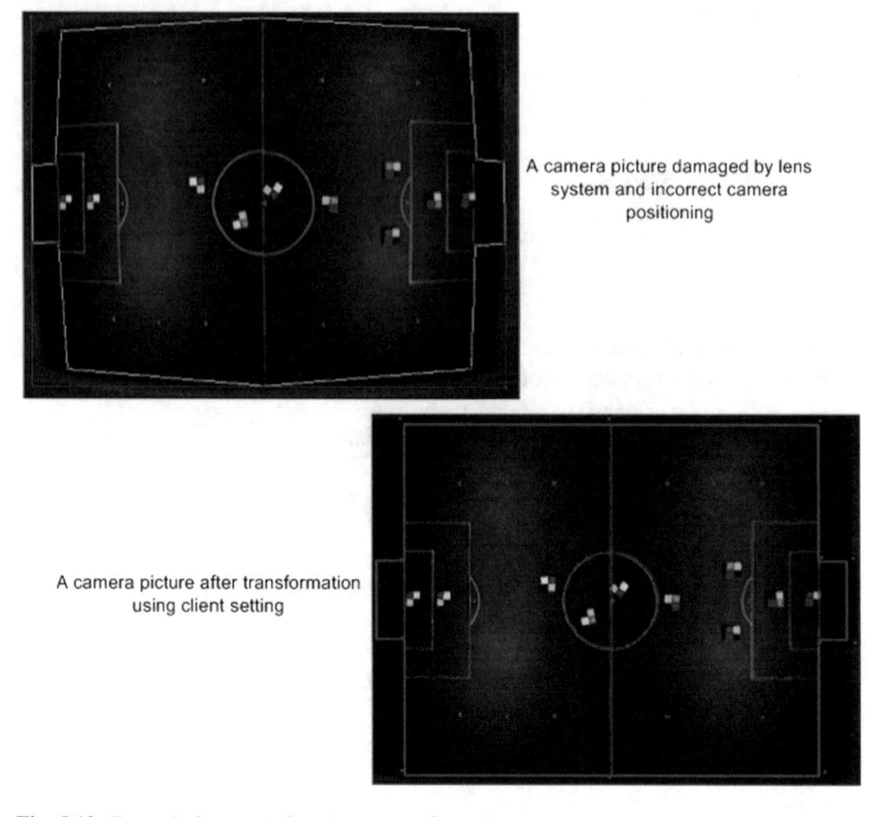

Fig. 5.13 Frame before and after picture transformation

5.4 Elementary Agent Players

They are software modules which are capable of interference with the game after robots have been allocated to them and they are trying to solve subtasks they have been given. In the developmental process there were 12 elementary agent players with the following differences:

- the area of operation
- planning and motion tracking
- behaviour on the field towards the position of the opponent and the ball.

By the process of allocation of the robots to the agent players, the complex agents are created which are active on the field and capable to dynamically affect the game. The detailed description of these agents can be found in Chap. 7.

5.5 Communication Module

Programming module which processes data (speed and rotation speed) and match them with agent players. After data chaining has been performed, these are then sent out of the computer via USB interface and later received by data senders. This module appears to be the simplest part of the program concerning the complexity and code size. The module code is included in one of the functions (procedure).

5.6 USB to RS485 Converter

Hardware module which transforms data from USB to RS485 interface. The only condition is that the bus 485 has no problem with data transformation as fast as 921.6 Kbit/s. The wiring diagram as it was designed and implemented is displayed in the Fig. 5.14.

5.7 The Transmitter

Projected communication system between robots and computer is composed of five transmitting modules with the basic frequency 2.4 GHz. Projected and executed wiring diagram can be seen in the Fig. 5.15.

Each module has chosen its own line data channel (speed and rotation speed) to transmit data cyclically to one robot. The data transmission speed is 921.6 Bkit/s. The cycle lasts for approximately 1 ms at two 2-bite data transmission, auxiliary data transmission, and transmission service. Repeated data transmission lowers transmission defectiveness. Transmitted data will change after the new data chain is intercepted.

5.8 The Robot Player

The object capable of the action based on three main units:

- Structural hardware: chassis, side plates, wheels, train of gears.
- Electronic hardware: DPS with embedded master microcontroller, transmitter and H-bridge to control the drives.
- Control software: processing of received data, their conversion to drives speed, and PSD performance of regulator of drives.

In the process of development there has already been the fourth generation of chassis used which is 4-wheeled (Fig. 5.16).

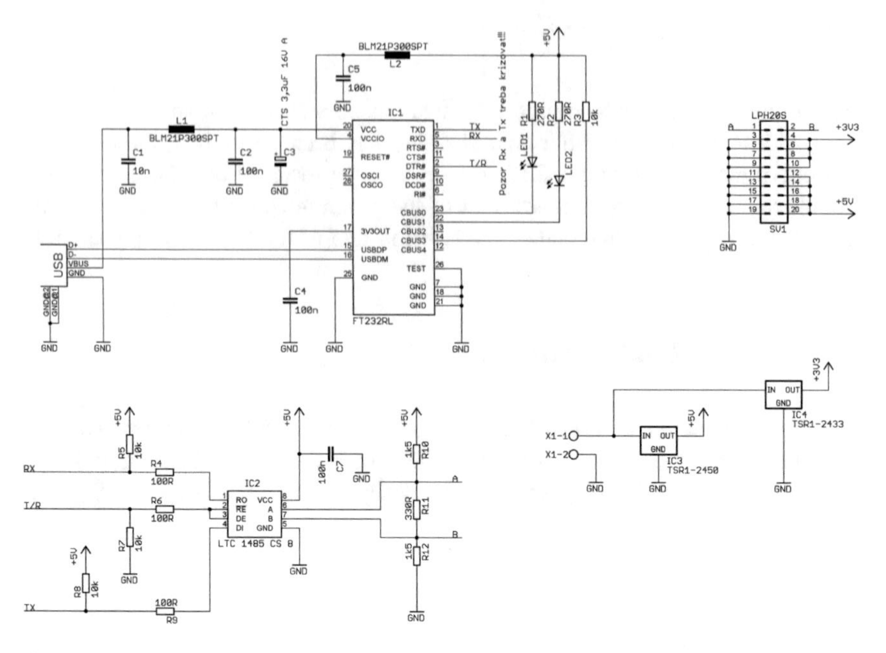

Fig. 5.14 USB (FTDI) to RS485 converter plug in scheme

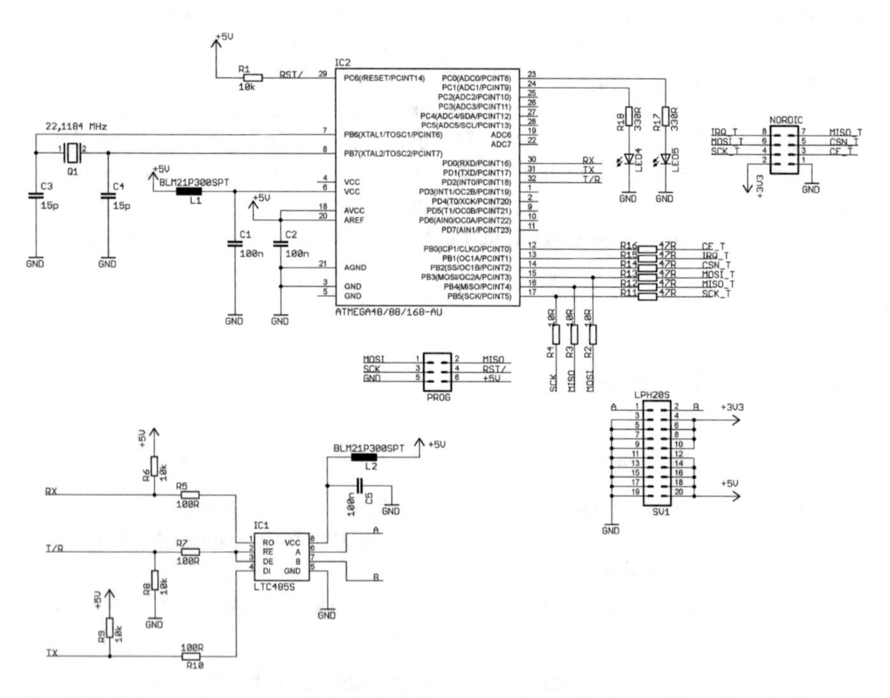

Fig. 5.15 Scheme of output module for wireless transmitter with data sniffing for RS485

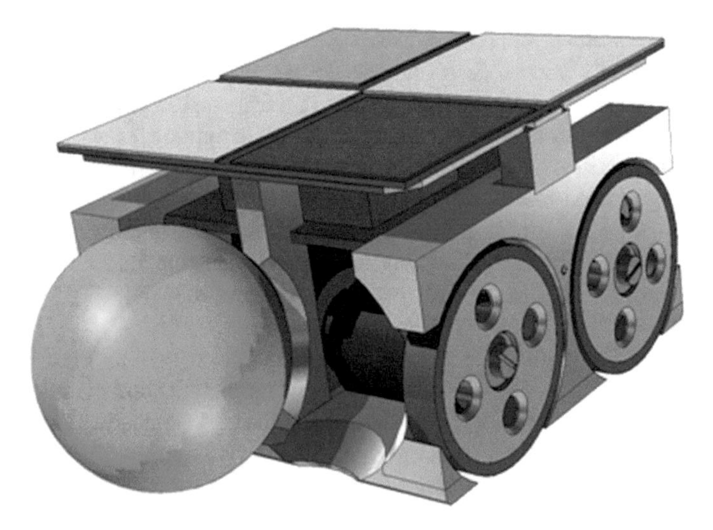

Fig. 5.16 Structural concept of the last version of robot players [50]

The most important construction request was to move the robot's centroid as low as possible while maintaining its central location. The electronics design and DSP set up with the size 54 × 55 mm were derived from the same request. Block diagram of electronic hardware with supply and signals can be seen in the Fig. 5.17.

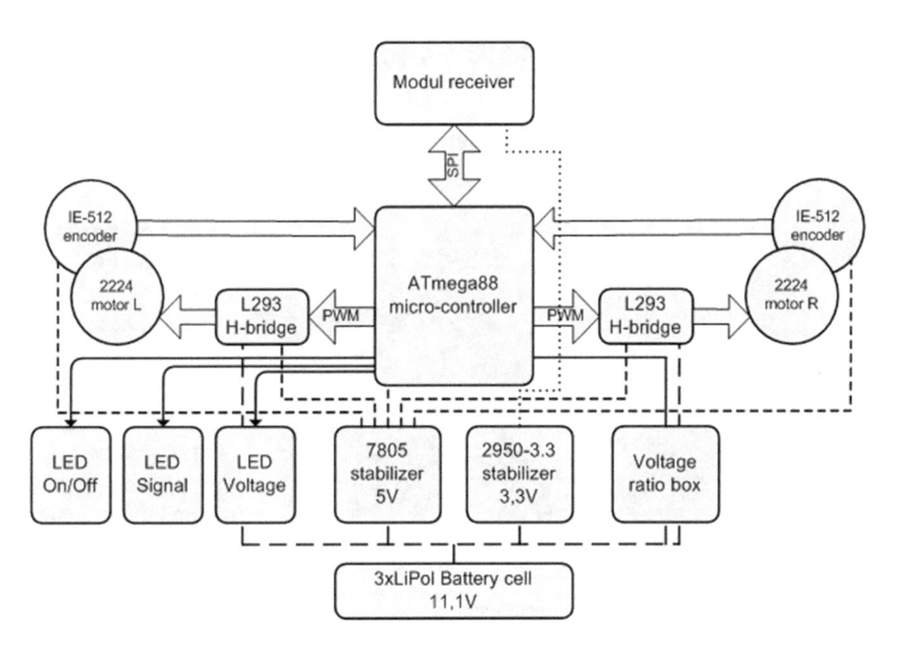

Fig. 5.17 Block scheme of robot hardware

Minimum external parts were used out of microcontroller in the electronic design. The robot's "Brain" is ATmega88-16AU. Its driving voltage was changed into 22.11 MHz to be capable of achieving basic tasks: communication with receiving module through SPI 5 MHz, impulse computing as well as encoded sequence of impulses, software PSD regulator with sampling 1 ms, PWM outputs for drive regulation, speed and angular velocity transformation to desired wheel speed, voltage power supply control. In addition, the used microcontroller did not show any defects after experimenting tests and driving voltage have been used during the communication via SPI with the receiving module with input voltage 3.3 V. DPS design in Eagle program can be seen in the Fig. 5.18.

DPS is supplied with three analog integrated circuits, two stabilizers, and two connectors for ISP (In system Programming) and communication module 2.4 GHz. Software flowchart performing drive control and the communication with receiving module is displayed in the Fig. 5.19.

The body of the main program (a) begins with microcontroller initiation. Inputs and outputs are set up, together with interruptions, PWM, SPI. Circle monitoring of IRQ flag follows, which is set up at the moment the data are received. After the received data have been loaded, the waiting process of data receiving which is cyclically repeated follows. The interruption of build up edge of output from encoder (c) has the same process for both encoders (left and right). After the interruption has occurred, the second encoder output is being controlled (EXT_ENC2). The number of impulses indicating speed and sense of engine rotation is incremented or decremented in accordance with its state. Interruption

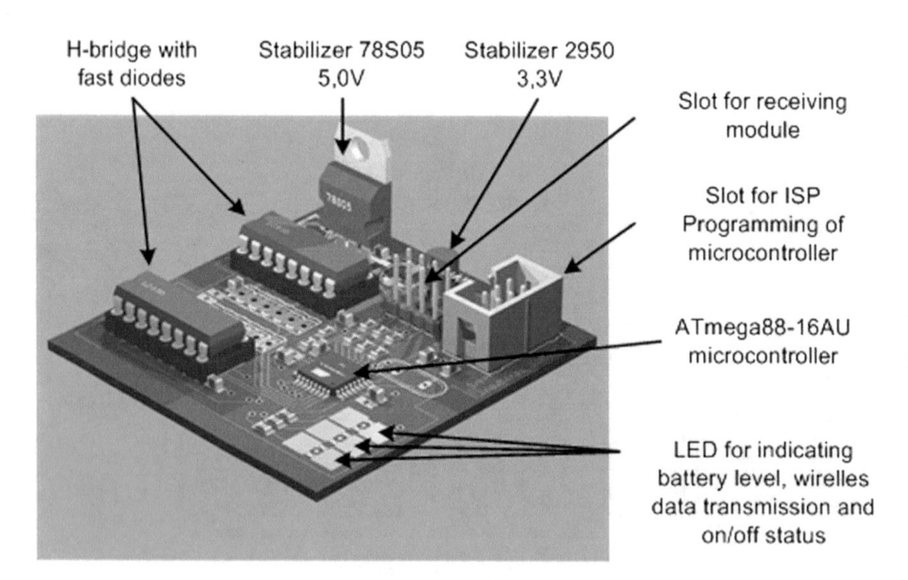

Fig. 5.18 DPS designed for robot electronics

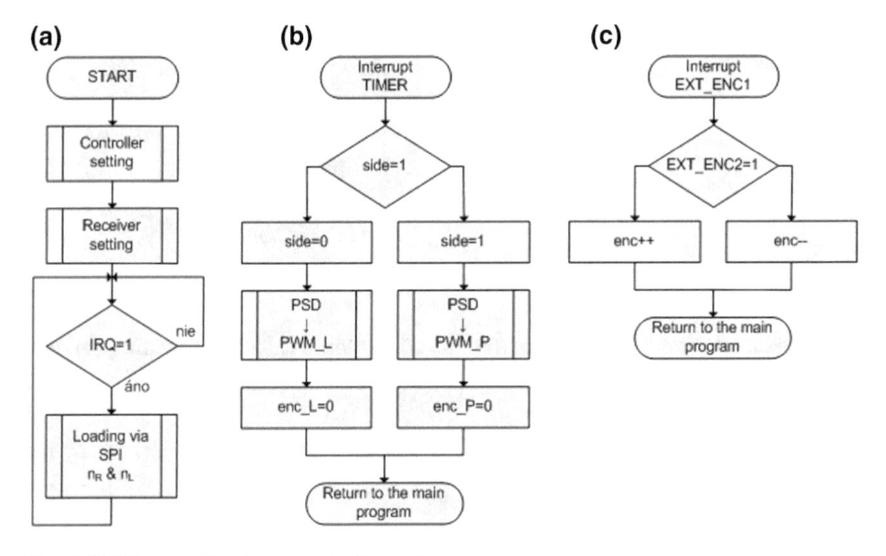

Fig. 5.19 Diagram development of robot software

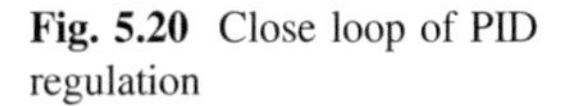

Fig. 5.20 Close loop of PID regulation

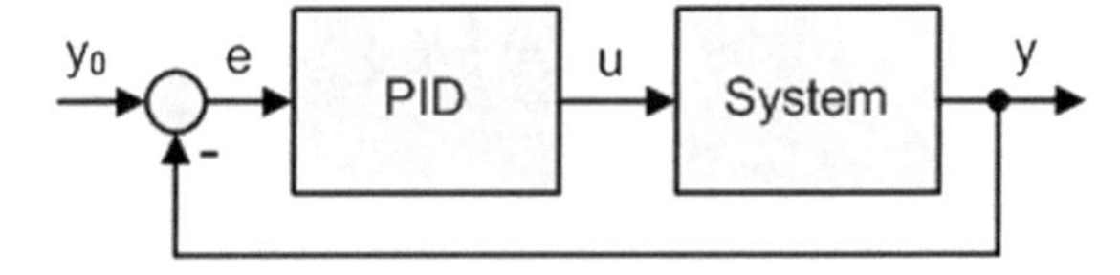

from timer (b) occurs every 500 μs. In the beginning, the PSD computing result is separate for left and right engine by means of variable side. Strobe frequency for both PSD controls is 1 ms whereas their operation is shifted in 180°.

Application note is used for computing [80]. Continuous loop with PID controller can be seen in the Fig. 5.20.

PID controller compares measured value y with desired value y0. The difference is error e entering the regulator. The regulator thus corrects control variable u which enters the regulated system. PID controller is composed of three components:

- Proportional
- Integral
- Derivative (D-controlller).

Block diagram is in the Fig. 5.21.

System transfer function (Fig. 5.21).

$$\frac{u}{e}(s) = H(s) = Kp\left(1 + \frac{1}{T_i s} + T_d s\right) \tag{5.9}$$

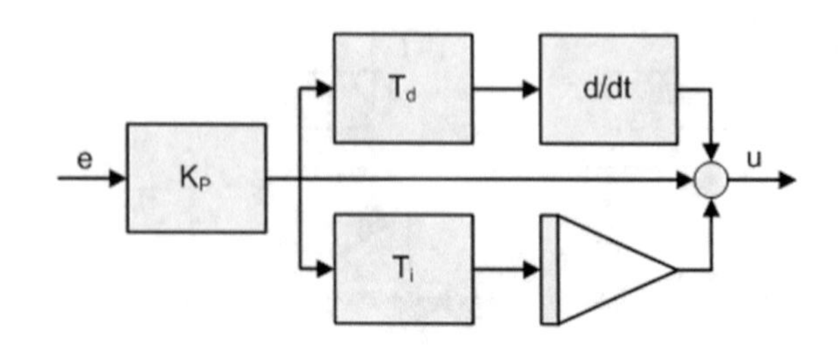

Fig. 5.21 Block scheme of PID regulator [80]

The general form of equation of the PID controller to calculate actuating variable u is [80].

$$u(t) = Kp\left(e(t) + \frac{1}{T_i}\int_0^t e(\sigma)d\sigma + Td\frac{de(t)}{dt}\right) \tag{5.10}$$

After adjustments for the PSD controller which is the discrete form of analog PID controller, this equation looks as follows [80].

$$u(n) = K_p e(n) + K_i \sum_{k=0}^{n} e(k) + K_d(e(n) - e(n-1)) \tag{5.11}$$

whereas n represents the number of patterns. The Summation term had to be limited in the application and its reasonable maximum had to be found for speed control and to avoid overflowing of the summarized samples in the given variable.

5.9 The Application Control Software

The development of software part within the driver consisted of three main modules (Fig. 5.22).

Image processing module: output module represents positional data of all the objects participating in the game (10 robots, the ball).

Strategic module: was formed out of 6 individual units specifically, the main agent —master and 5 agent players. The goal of the agent master is to make decisions on team players strategy. By the process of joining software modules (players) with their "physical body—robot", agents (robot players) are created, they are able to affect the situation on the field (Fig. 5.1).

Communication module: the simplest module regarding the number of code lines. It provides the collecting and transformation of all robot commands to the transmission chain. The communication module also provides transfer of transmission chain to transmitting modules.

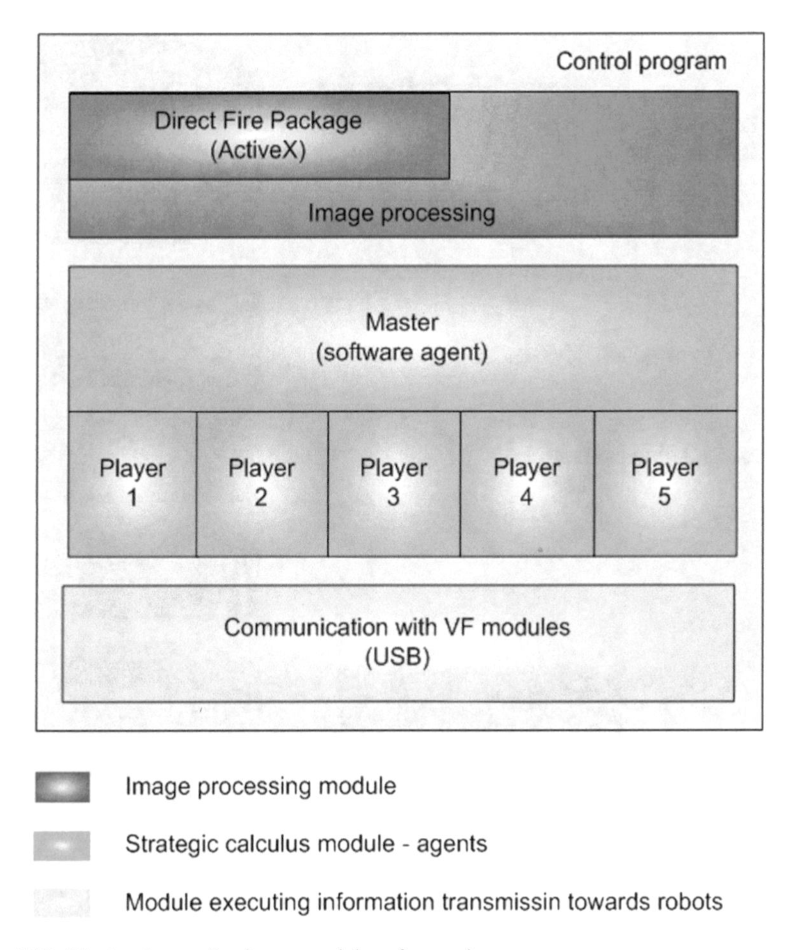

Fig. 5.22 Block scheme of software modules of control program

Figure 5.23 displays the functional blocks of the driver with the performance sequence in vertical mode and information flow being marked. Each camera frame is firstly saved in cache memory. The frame of size 568 $\times$ 426 pixels is saved in three 8-bite fields which are represented by primary colours (R—red, G—green, B—blue) ActiveX element provides the indices to the beginnings of these fields. After the frame has been saved, the identification and separation of all the colours take place which are necessary in order to distinguish the robots and the ball. The next step is the identification of the clustering colours and computing of theirs centres coordinates out of which robots coordinates in plane (Rn—robots, Rsn— opponent's robots) and the angle of rotation towards the basic coordinate system (φ) and the ball coordinates (L—ball) are computed. The parameters defining these data are saved in time sequence for 67 last captured frames i.e. an archive of motion from the last

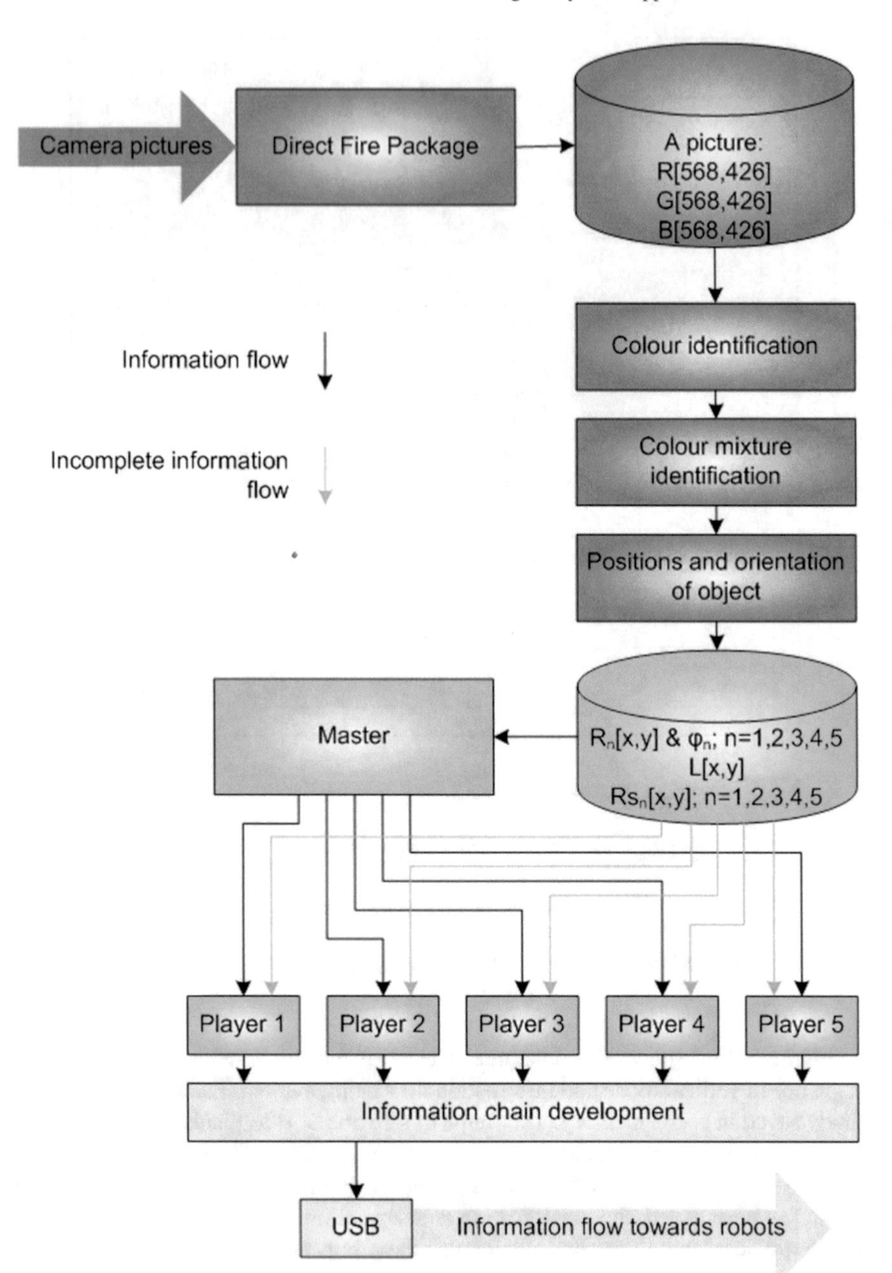

Fig. 5.23 Functional blocks of modules of control program

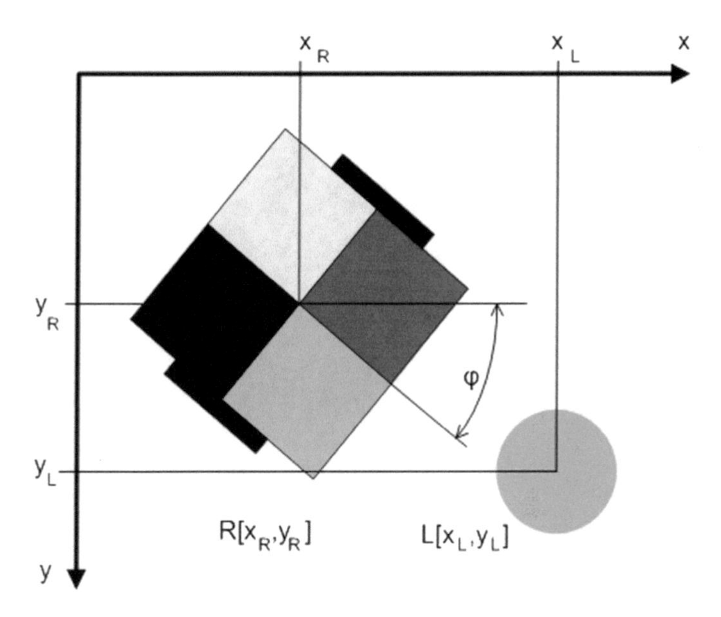

Fig. 5.24 Parameters determining absolute object position

second is being created. This archive is used by agents to predict next step of both the ball's and opponents' movements. The agent master decides on the chosen strategic action in accordance with chosen type of the basic strategy. This information is further transmitted by form of messages to each agent player. Players must cooperate and make compromises not to collide among each other and in order to follow all the strategic instructions. Finally, the generated motion information is processed to the transmission chain which is transferred to transmitting modules via USB interface (Fig. 5.24).

Chapter 6
Hierarchical Arrangement of the Agents in Developed MAS and Its Strategic Concept

This chapter deals with the descriptions of hierarchical arrangement of the agents in designed system and relations among agents within their strategic approach towards general strategy of system behaviour. The design of the agents, their hierarchical arrangement and force distribution to general strategy among agents was derived from the main application goal:

- To achieve higher score than competitor

The main goal was divided into two basic goals:

- To score the goal
- To prevent the opponent from scoring the goal.

6.1 The Agents and Their Tasks

Before the hierarchical arrangement design was completed, it has been considered that basic strategy of robots' behavior is to be chosen by the user of the driver. They can choose from the predefined defensive or offensive strategies. Their options of affecting players' position are wide, however the ability to choose a strategy is limited by the number of predefined strategies. The general strategies choice is the choice of the highest level and can be performed only during the interruption of the game or pause. Strategic behavior of the system throughout the game as a dynamic reaction to the aroused situations is on the intermediate level. Behaviour of the system during the game is affected by the chosen strategy on the highest level. Eventually, the strategic behavior of each member team (robot-player) in on the lowest level. The individual strategic levels with the illustration of size and importance of affecting strategic behavior of players are shown in the Fig. 6.1.

M. Hajduk et al., *Cognitive Multi-agent Systems*, Studies in Systems, Decision and Control 138, https://doi.org/10.1007/978-3-319-93687-1_6

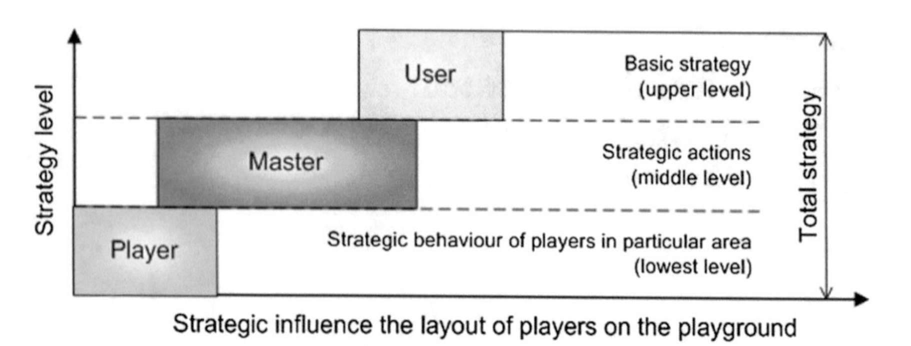

Fig. 6.1 The illustration of the options to affect the strategic positions on the field

Hierarchical arrangement of the agents was designed and implemented for described strategic concept. It may be seen in the Fig. 6.2, where the master takes over strategy control on the intermediate level and the agent players on the lowest level.

MAS is composed of 6 agents, whereas each agent is superior. After the user has chosen the strategy, the master chooses the suitable strategic action to achieve the goal. It is possible for the master to choose 5 agents out of the elementary agents set (Chap. 7) whenever throughout the game that are available for the chosen strategy. The master later matches them to individual robots pursuant to the situation and position of the robots on the field. Each elementary agent is the agent which performs some subtasks in order to accomplish the main goal. The main goal of multi-agent system is to reach higher score than the opponent. This main goal is divided into two basic goals: to prevent the opponent from scoring and to score goal. Both goals are divided into elementary goals with certain sequence and progression by each agent. The elementary goals always encourage the agents to achieve the highest goal.

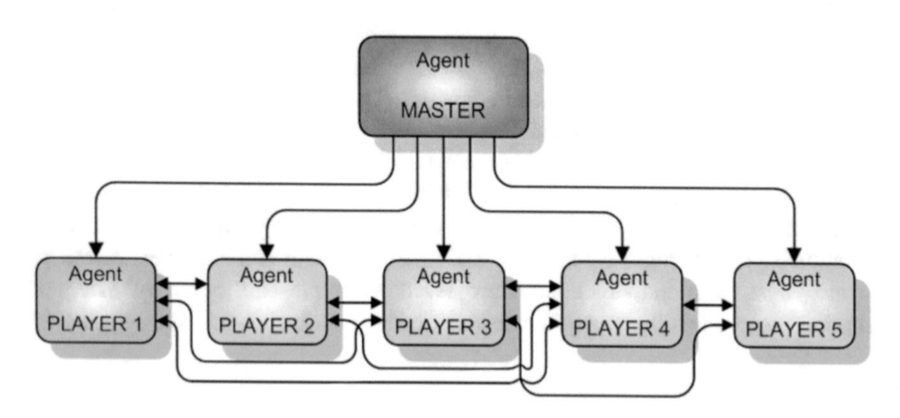

Fig. 6.2 Hierarchic configuration of used MAS

Since the master can change players' position throughout the game, the chance to choose the most suitable action was on the lowest level and agents' complexity problem was solved by creating many elementary agent players. Their existence is defined by the superior agent master (Chaps. 5 and 7). It is responsible for the selection from the actions: absolute defensive, defensive, midfield game, offence, absolute offence (Fig. 6.10). The individual actions are chosen based on players' position, and their current speed and accelerations, the position of the ball, prediction of its movement, and prediction of opposing team's movement. Chosen overall strategy (different levels of offensive and defensive strategy) affects the set of agents which are available for each action required. Eventually, the decision is made to allocate ID to each elementary agent player following the current position and direction of motion of its robots. When decision is being made (of existence and ID allocation), the master takes into account the previous state of agents.

6.2 Basic User Strategies Implemented in the Driver

GUI of the driver has specified area (red dash line in the Fig. 6.3) in the main window in which the basic strategies are set up by the user on the highest level by means of 4 option combination. After the games has been initiated by the user, the desired redefining of agent's (master) parameters takes place in the driver. Consequently, the agent's (master) behavior will depend on the these settings throughout the game.

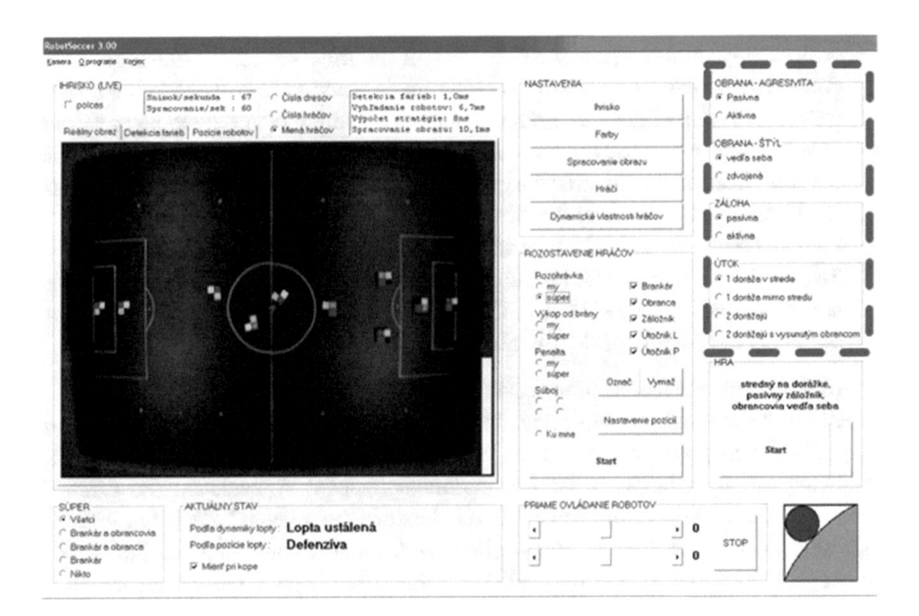

Fig. 6.3 GUI of control program with particular area of strategy settings

The affect of basic strategy lies in the following 4 options and their possibilities:

(a) DEFENSE—AGGRESIVITY defines the behavior of defenders when defending without the ball.
Passive: Agents' (defenders) behavior is set up to prevent defenders from being active in blocking movements of the opponent i.e. not creating any obstacles when moving towards defense line.
Active: Defenders' behavior is set in such a way not to be able to actively stand in opposing team's course when playing without the ball. The agent checks this part of its behavior to prevent from breaking rules.
(b) DEFENSIVE—STYLE defines the type of selected pair of defenders depending on the position of defense game. Technically, the agent master changes its decision-making how to employ one out of two pairs of elementary agents.
Side by side: Agents defenders will be positioned in front of their defense zone side by side during the process of defense. Each defender performs its actions on its side, i.e. they will be divided into right and left backs.
One after another: Agents defenders will double their position and both of them will perform their actions throughout the whole field length (full-back back and midfield).
(c) BACK-UP defines the player's behavior in the centre of the field
Passive or active: agents' behavior as it is with DEFENSE—AGGRESIVITY
(d) ATTACK defines how many strikers will support the defender that actively watches the ball and in what position strikers will play:
One striker rebounds the ball in the centre: There will be two players attacking, one of them will change its position in the centre in order to score the ball in the desired way. Active striker watches the ball and tries to seize it. When the striker possesses the ball, the decision is being made whether to strike it to the goal or passes it in front of goal. The other 3 players remain in their defending positions.
One forward rebounds the ball out of the centre: Similar to previous case with the modification where striker (right or left) waits for rebounding of the ball on the opposite side towards the position of active striker.
Two strikers rebound: Three strikers attack (active, right and left).
Two strikers rebound with the central midfielder: Similar to previous case, three strikers attack, supported by the central midfielder as a back-up. Only goal-keeper remains in the offence.

All the combinations of these basic strategy settings are in the Fig. 6.4.

All the basic strategies are derived from two basic ones: 2 cooperating strikers, three cooperating strikers. The user might choose out of 24 combinations. The following strategic behavior on the intermediate level depends on the master's decision. Behaviour on the lowest level depends on players' decision. Graphic representation in the Fig. 6.1 shows that even players can influence the general strategy of the system. This possibility is limited to player's movement within its area of operation.

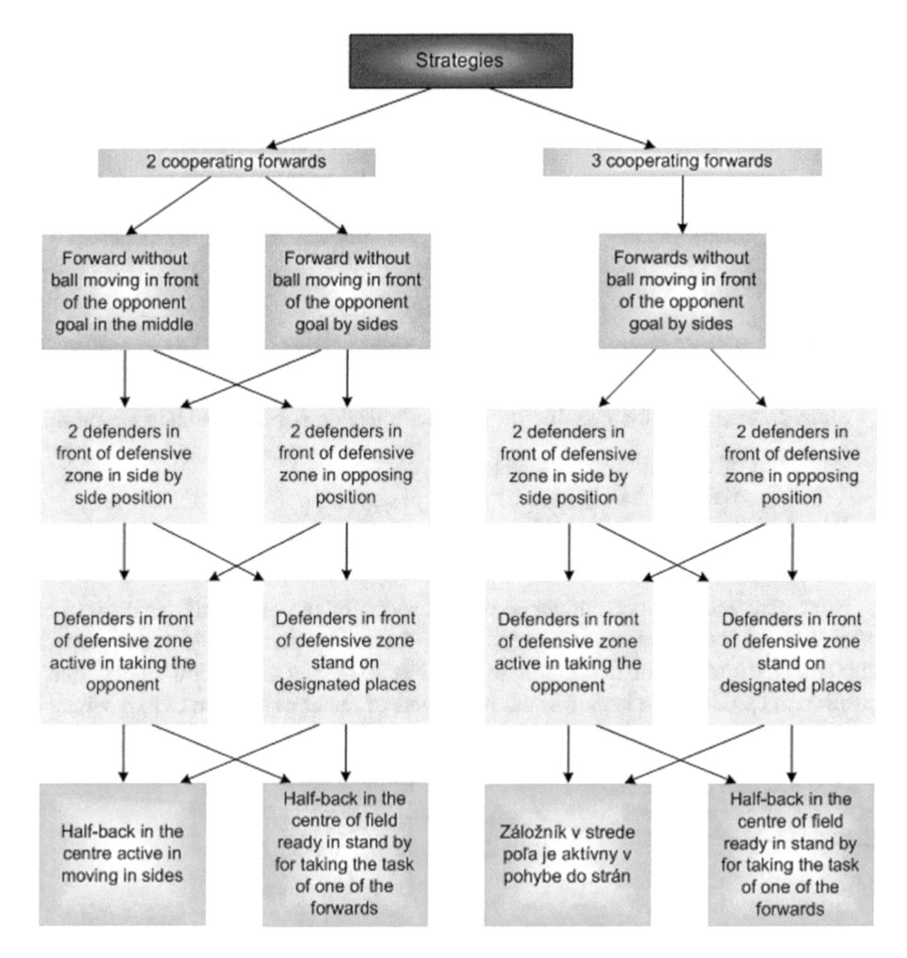

Fig. 6.4 Combination of used strategies on top level

6.3 The Selection of Strategic Actions by the Superior Agent

The possibility of intervention and strategy change results from the rules MiroSot category [67] and is limited only to interruption of the game by the referee. Unlike user, the master decides on the strategic action, after each processed frame during the game. Decisions are affected by the way basic strategy of the user is set and instant situation on the field. Instant situation results from speed and position parameters of the ball (Fig. 6.5). Position parameters are: absolute position of the ball on the field and relative positions of the ball with reference to robots on the field.

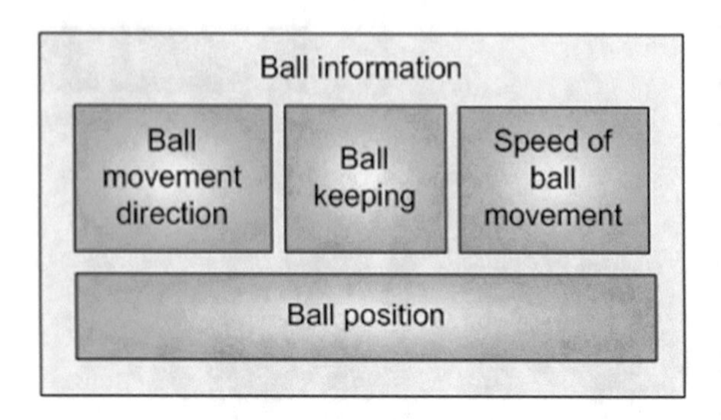

Fig. 6.5 Block scheme of the main parameters influencing master decision making on immediate strategy action

Main parameters which enable the master to decide for the strategic action are:

- Direction of the ball's motion (Fig. 6.6)
- Who possesses the ball (Fig. 6.7)
- Speed of the ball's motion (Fig. 6.8)
- Position of the ball (Fig. 6.9).

Next parameters which influence the decisions of the agent master depend on prediction of the situation in a very close future. In performing agent there is implemented prediction maximum in 1 s. For example: prediction of the ball's position in case it's free is performed my means of linear regression of axis where it

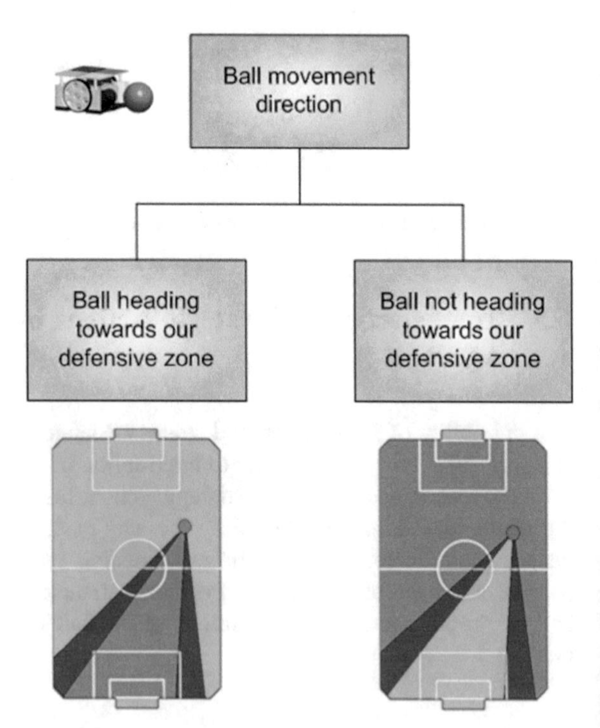

Fig. 6.6 Decision of agent influenced by ball movement direction

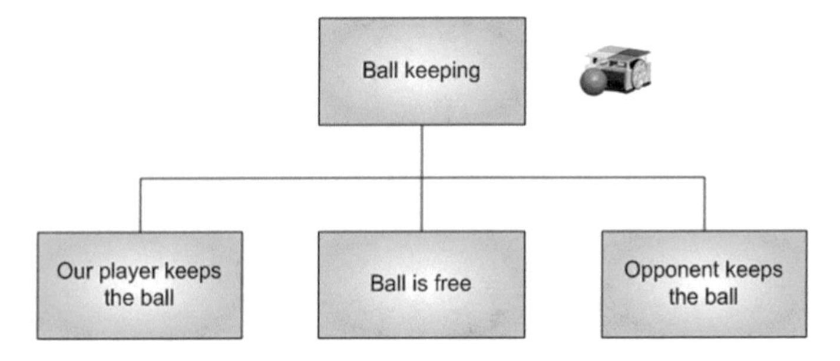

Fig. 6.7 Agent decisions influenced by ball keeping information

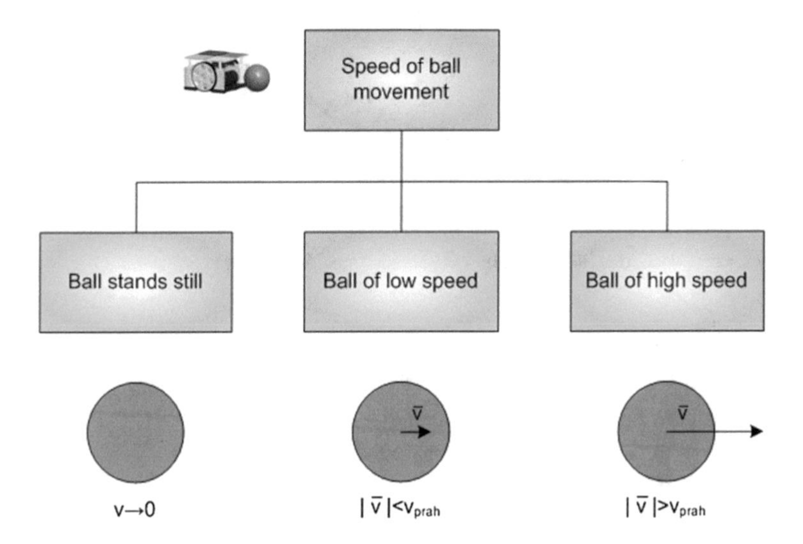

Fig. 6.8 Agent decisions influenced by speed of ball

does not allow for deceleration of the movement in such a short time period. The agent uses the rule of the same angle of reflection as the angle of impact to side walls of the field.

The direction of ball's movement is defined by two states (Fig. 6.6).

- the ball directs at our defense line.
- the ball is not directed to our defense zone (this assumption is also true for the ball without any movement.

Figure 6.6 shows light red colour which displays the area of the true sentence stated in the block over the field diagram. Decisions have to be stable without oscillation among frames. This is the reason why the hysteresis has been applied to thresholds of switching. It can be seen in dark red colour in the picture. Its width provided the stable switching of states not affected by the oscillation of position among frames.

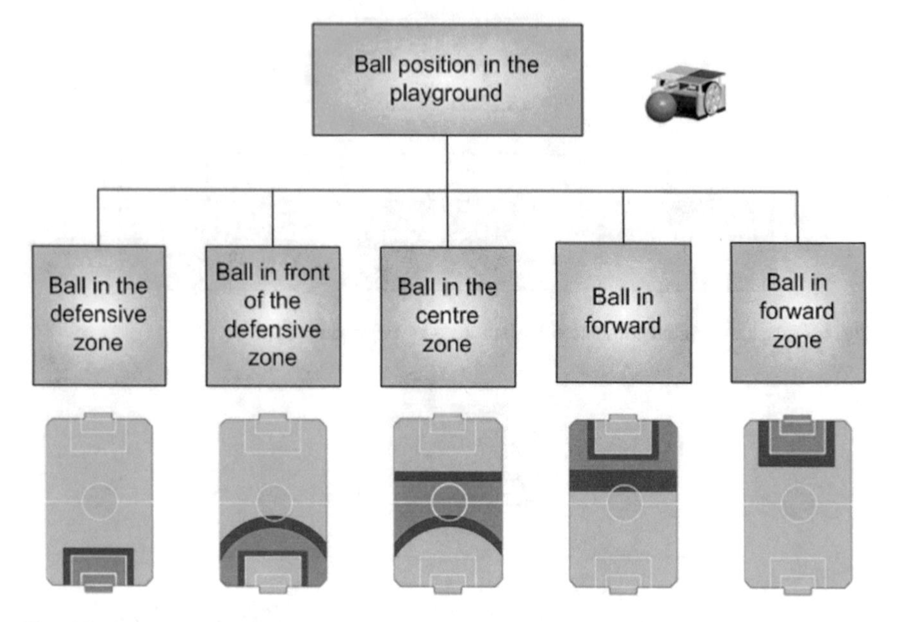

Fig. 6.9 Agent decisions influenced by ball location/position in the playground

Sentence related to the possession of the ball (who has got the ball under control) has three solutions: (**Chyba! Nenašiel sa žiaden zdroj odkazov**.):

- The ball is in the possession of our player
- The ball is in the possession of the opposing player
- The ball is free.

The possession of the ball by any of the robots is defined as threshold distance of the ball from the front part of the robot and width of the robot. These two parameters define the area in front of the robot and when the ball is to be found there, the robot has it under control. Hysteresis has to be applied here because of instability of coordinates of all the objects after the frame has been processed.

Speed of ball's movement offers three possible states (Fig. 6.8).

- Ball is in rest (speed approaching zero)
- Ball has slow speed
- Ball has fast speed.

Above mentioned parameters and prediction of the next situation affect the decision-making of the agent master. He is the one to moves the threshold values of switching among actions as it is in the Fig. 6.9, where red color means the area of ball's position.

In order to achieve stable switching of states, the agent operates with chosen widths of hysteresis which were defined in decision-making rules during the development of control software. The master decides on instant strategic action for

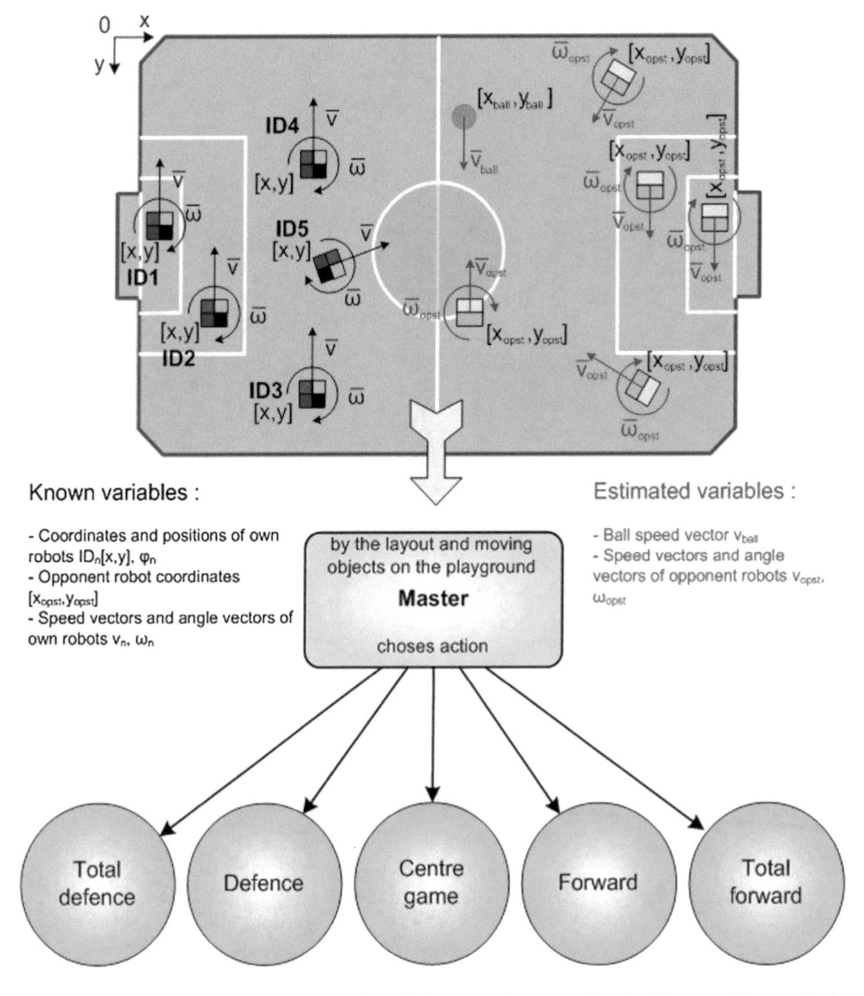

Fig. 6.10 Variables in the system which influence the master's decision making and the immediate action out of all defined actions

the basic strategy chosen by the user and selects out of five strategic actions which are shown in the Fig. 6.10.

Master's decision is made on the basis of the following pieces of information:

- • Ball's position $(L[x_{ball}, y_{ball}])$
- position and orientation of players $(IDn[x, y], \varphi n)$ and their speed parameters (vn, ωn), n = 1, 2, 3, 4, 5
- position of opponents' players Rsn [xopst, yopst], n = 1, 2, 3, 4, 5.

Individual strategic actions diffuse among each other in the sequence and skipping of actions should never happen. The only exception is faulty vision and subsequent image processing with brought large error.

Chapter 7
Elementary Agents Players and Their Assignment to Robots

It is obvious that the master decides on the basic strategic positions of the robots on the field based on the combination of hierarchical arrangement and above mentioned decision-making process of the strategic action. Set of all elementary agents which were crated is in the Fig. 7.1.

Set notation:

$$\mathrm{AH} = \{\mathrm{g, d, lb, rb, bd, cb, m, dm, lf, rf, cf, s}\} \tag{7.1}$$

g-goalkeeper, *d*-defener, *lb*-left back, *rb*-right back, *bd*-back defender, *cb*-centre back, *m*-midfield, *dm*-defensive midfield, *lf*-left forward, *rf*-right forward, *cf*-centre forward, *d*-defender.

Set of used elementary agents which is the subset of the set of all elementary agents is defined for each strategy chosen by the user.

$$\forall\, s\_high \in St \exists AH_{high} \subset AH;\ \left|AH_{high}\right| \geq 5 \tag{7.2}$$

s_high—strategy on the highest level.

St—set of created strategies St = {24 combinations of strategies}.

Set of five elementary agents is defined for each strategic action based on the chosen strategy, which is a subset of set of elementary agents for the given strategy:

$$\forall s\_mid \in Sa \exists AH_{mid} \subset AH_{high}; |AH_{mid}| = 5 \tag{7.3}$$

s_mid—strategic action on the intermediate level.

S_a—set of strategic actions S_a = {absolute defense, defense, midfield play, attack, absolute attack}.

The master assigns each elementary agent to one robot from the field after the decision has been made on strategic action and its defining action:

M. Hajduk et al., *Cognitive Multi-agent Systems*, Studies in Systems, Decision and Control 138, https://doi.org/10.1007/978-3-319-93687-1_7

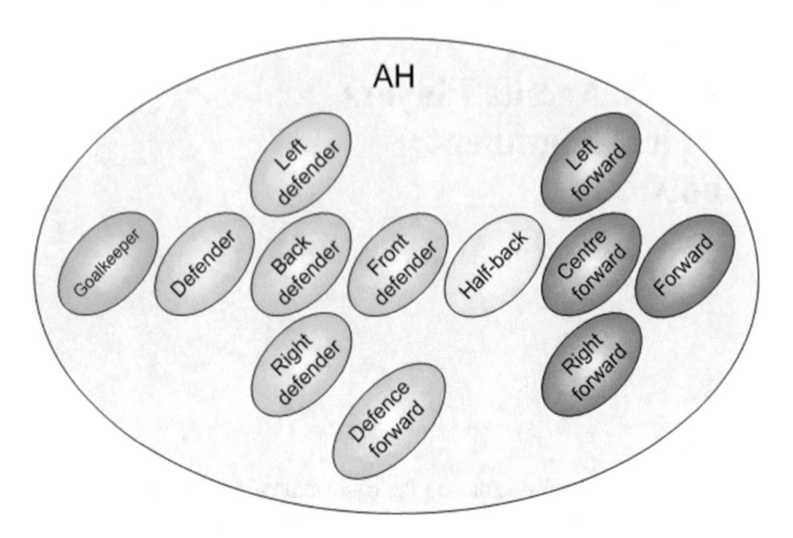

Fig. 7.1 A set of all elementary agents

$$\forall a \in AH_{mid} \exists! ID = a \tag{7.4}$$

ID-robot
a—elementary agent player

The agent master must assign each ID on the field to some of the elementary agents players and it after makes the decisions within its strategic position autonomously in cooperation with the others by means of assignment rules. All these agents are allowed to perform their actions within the playing area which was allocated to them during the development of software. Within its area of performance it controls its movement in order to prevent from collisions among robots and to achieve its elementary goal.

7.1 The Goal Keeper Agent

The basic rules of the agent:

- To stand on the line defined by vector of the ball's speed heading to the goal with its physical center of the front side and thus preventing the opposing team from scoring by defending the goal.
- To try to release the ball at rest or at very low speed out of defense zone. The area the agent may act within the field is shown in the Fig. 7.2.

Red color defines the area in which the agent is allowed to operate. Every time it appears to be out of this area, its priority is to get into this area. Line segment defines the priority track of basic movement. The agent comes out of this line only in case the ball needs to be brought out of defense zone.

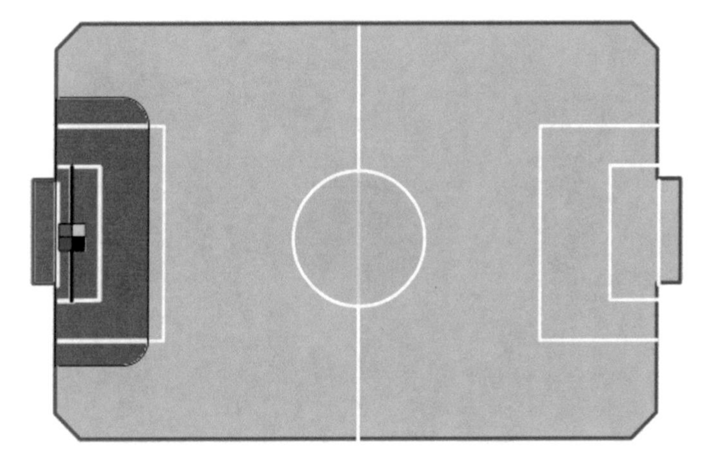

Fig. 7.2 Goalkeeper area of operation

Such an agent is utilized 100% in each strategy, i.e. there is no play without the goalkeeper. Even at maximum pressing, this possibility would be too risky, since it is likely to concede the goal as it is e.g. at real hockey match.

7.2 The Agent—Defender

The basic roles of the agent:

- To stand on the line defined by vector of the ball's speed heading to the goal with its physical center of the front side and thus preventing the opposing team from scoring by defending the goal.
- To try to release the ball at rest or at very low speed out of defense zone and the area close to defense zone.

Figure 7.3 shows the area of operation of the defender and its priority movement to achieve goals. The agent has to be careful about not breaking the rules i.e. not to enter the goal zone intentionally where only one robot might appear at the process of defense. Its goal is to receive message from the goalkeeper. The goalkeeper might ask the robot not to block at kick-off. In case the goalkeeper performs kick off, the defender must leave the space by getting to the side. If the agent appears out of the permitted area, its main priority is to come back to the defined area. This agent is utilized 100% almost at all the actions apart from attack and absolute attack when following strategy with 3 strikers and one midfielder.

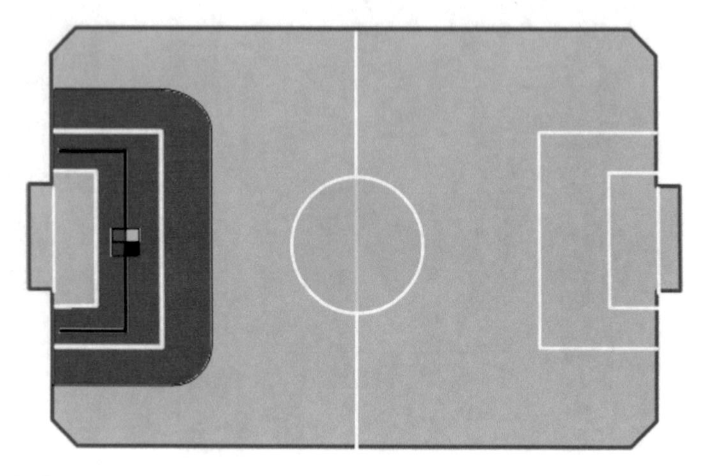

Fig. 7.3 Defender area of operation

7.3 The Agent—Right or Left Back (Defender)

The basic roles of agents:

- To stand on the line defined by vector of the ball's speed heading to the goal with its physical center of the front side and thus preventing the opposing team from scoring by defending the goal.

Figure 7.4 shows the area of operation of right back. Left back has the same area of operation as right back, at the reflection across vertical axis of the field. The agent, from the point of view of its movement is rather simple but its algorithmical compexity increases with the number of conditions it has to meet throughout the game and has to be alert not to break the rules. All these conditions affect its final behavior. If the ball is behind them, they take the most advanced position to get the ball when being pushed away from the defense zone.

In case of active defenders, they try to force the opponent to replan the track of players when rebounding from the space in front of the defense zone. They must "think" and not to break the rules by pushing the opponent out of the space in which it is already found, i.e. must force it to put through, not vice versa.

Right and left backs must cooperate in order to come to an agreement to occupy the central positions, since both players operate in the central zone as well. They must cooperate with both the defender and the goalkeeper which may kick off the ball from goal or defense zone. They must cooperate with attacker as well, which may tackle the gained ball out at back corners of the field.

Agents—right and left backs appear mainly at defensive actions within all the strategies and their ID is often swapped during the process of switching from defense to offense and vice versa.

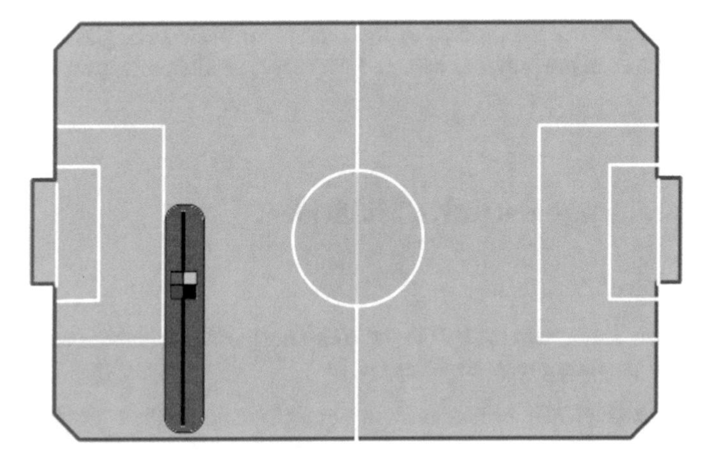

Fig. 7.4 Right defender area of operation

7.4 Wing-Back Agent or Centre-Back

The basic roles of the agents:

- To stand on the line defined by vector of the ball's speed heading to the goal with its physical center of the front side and thus preventing the opposing team from scoring by defending the goal.

Figure 7.5 shows the areas and priority movements of wing-back and centre-back. These defenders are simple regarding the movements, as it was in the previous case, however, regarding the final behaviour, their complexity rises rapidly. They must cooperate with almost all the players on the field because of the eventual collisions.

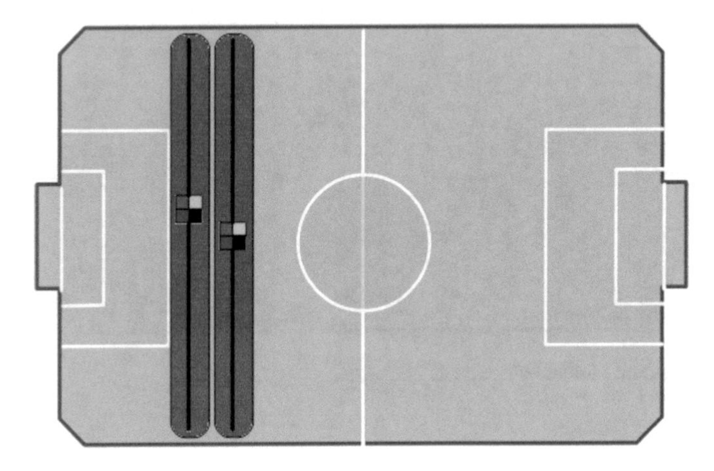

Fig. 7.5 The area of operation of wing-back and centre-back

They occur during the process of the actions for basic strategies in the similar manner as it was in the previous case and they serve as alternatives for right and left backs.

7.5 The Agent—Defensive Midfielder

The basic role of the agent:

- To wait until they win the ball in the space next to the defense zone in the most advanced position for its tackling.

This agent (Fig. 7.6) occurs only at strategic action called absolute defense. Before switching from defense, it used to be striker, trying to tackle the ball. Since this action is not always successful because of the opponent and the ball gets to defense zone, the master must use this type of a striker. Providing the master did not use the striker, the many rules would likely to be broken and more than two robots would appear in the defense zone. This commitment of offence would be awarded by penalty kick.

In case the ball gets in defense zone to the other side, it is expected by the agent, the rearrangement of defenders and this agent in front of the defense zone follows. The master evaluates when it is necessary to rearrange them and how to rearrange individual actions in order to make sure that the defensive midfielder occurs next to the other side of defense zone.

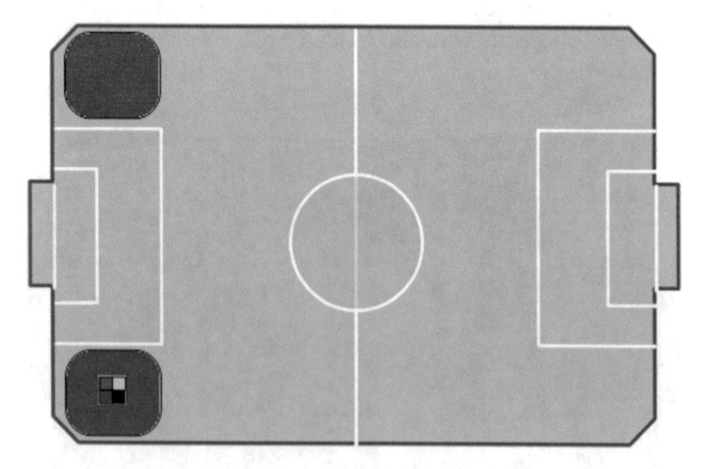

Fig. 7.6 Defensive midfielder's zone of operation

7.6 The Agent—Midfielder

The basic role of the agent:

- To wait until they win the ball in the centre circle in the most advanced position for its tackling or one of the strikers might be replaced by the midfielder.

The agent central midfielder (Fig. 7.7) was created as a type of the midfielder player. It is mostly used for strategic actions attack and absolute attack or at agents' transitioning from defense to attack. This agent may be active or passive as defenders. During transitioning from defense to attack it is often used by the master to stop fightbacks yet in the centre circle of the field. At the above mentioned process, the agent will wait for the opponent with the ball until the opponent gets behind the agent or until the direction of the ball is changed.

This agent occurs in all formed strategies like goalkeeper and defender. Unlike goalkeeper, the ID switches very often.

7.7 The Agent—Centre Forward

The basic role of the agent:

- To take the advanced position - nearest to the opposing team's goal in the centre, in order to be able to score the ball by the rebound.

It is type of a player which moves only in front of the opposing team's goal (Fig. 7.8). Its key task is to knock the ball down for a team mate to score. The ball might appear in front of the player in case the striker loses the ball and it is bounced

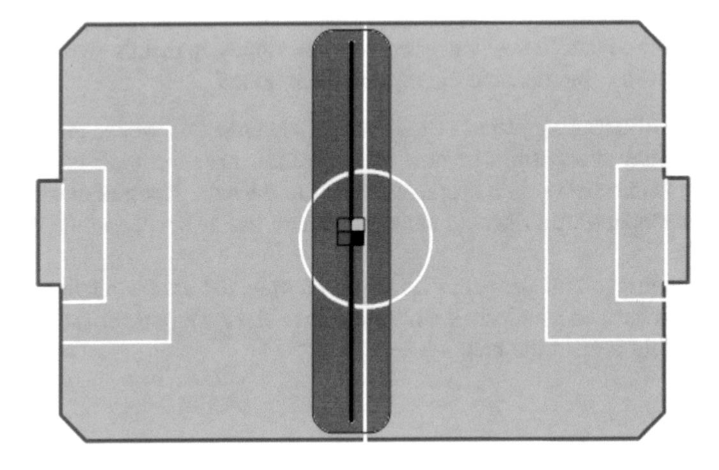

Fig. 7.7 Central midfielder's zone of operation

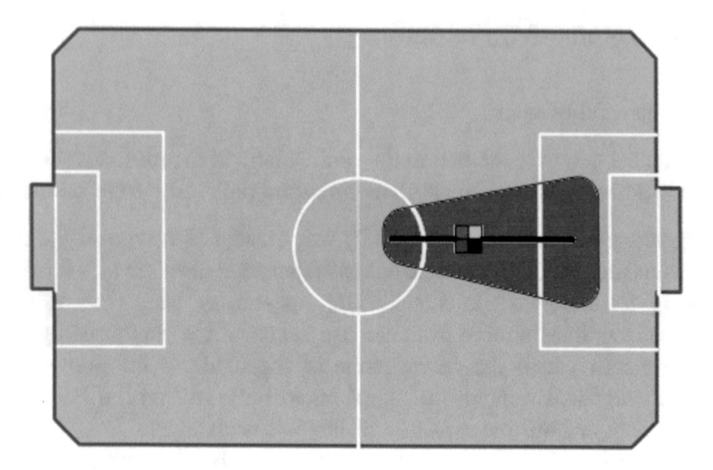

Fig. 7.8 Centre forward's zone of operation

back to the space in front of the goal, or in case of intentional pass from the striker. The secondary task of the player is to take position on the field within its zone in a way to be prepared for change the moving striker towards the ball anytime and "steal" it.

This type of the player is used for attack or absolute attack. Its exploitation at other action is not clear because of the position it takes which is on the halfway to the opponent.

7.8 The Agent—Right or Left Forward

The basic goal of the agent:

- To take the advanced position—nearest to the opposing team's goal on the right, in order to be able to score the ball by the rebound.

The type of attacking player called "winger" is used as it was mentioned above (Fig. 7.9). They are positioned in a wide position near the touchlines, it is the analogy of centre forwards movement slightly to the side. None of these strategies is used for rebounding strikers at once. Maximum two wingers and one striker are used.

Its usage throughout the game is like in the previous case explicitly connected with the actions such as attack and absolute attack as a support for striker with the ball or heading toward the ball.

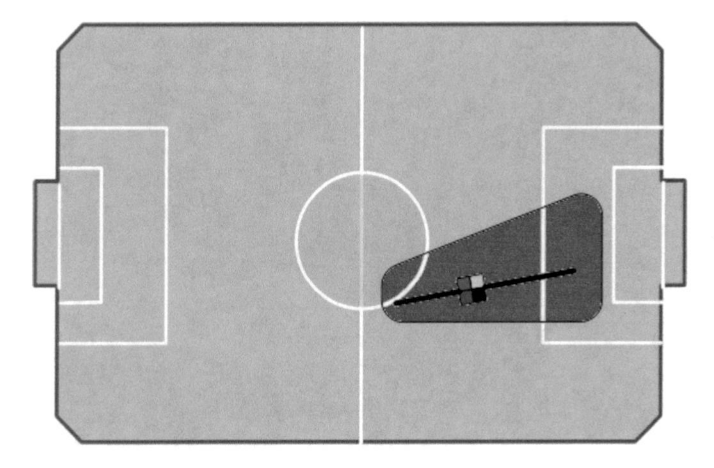

Fig. 7.9 Centre forward's zone of operation

7.9 The Agent—Striker

The basic goal of the agent:

- To receive the ball in the shortest time possible and to shoot it in the opponent's goal.

The most frequently used type of a player apart from the goalkeeper. The striker (Fig. 7.10) is the most complex regarding the algorithmical complexity, the most compound regarding the behavior and planning its movements. Deliberative part of the agent is on far higher level comparing to the other players. Its movement is not planned in line from point to point, but its plan is composed of two track types: straight-line motion and uniform circular motion. It must cooperate with the other players e.g. rebounding strikers. The centre forward performs passes to these players from the corners of the field based on their position in front or on the defense line of opposing team in case, that its move towards the goal is blocked by more opponents.

Striker is the most frequently relocated elementary agent regarding the rotation of positions. (change of ID assignment).

7.10 Motion of a Player Following Trajectory and Its Planning

The plan of players' moves is composed of two basic geometric shapes i.e. straight-line motion planning, and circular motion. The goalkeeper moves on vector parallel with the axis x and y and the end points at the corners edge of the goal most

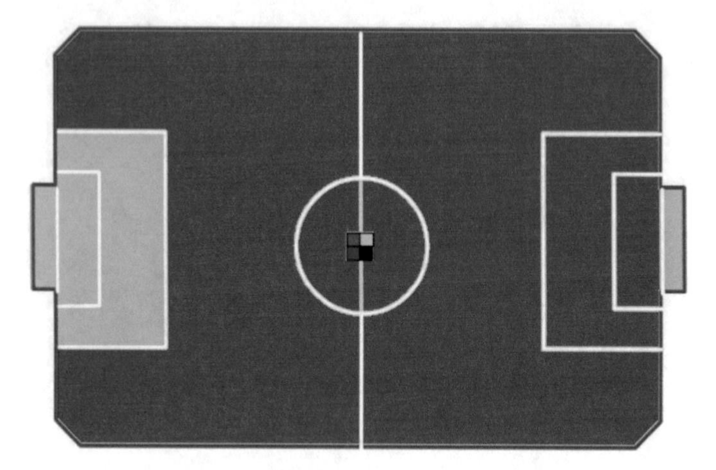

Fig. 7.10 Centre forward's zone of operation

of the time. Defenders line up in defending formations when performing defending tasks and their movement is composed of path following the section of a line. The player-striker is the most complex regarding the planning the path. Its motion consists of path planning on line ending with circle or just a part of a circle based on position of a player, the ball and opponent's goal. Motion on this path is achieved by computing of pixels the player is guided to. These pixels are computed from each frame. Movement of the players on the path seems to be continuous without rapid changes of direction for an observer (person).

The way the player's movement is generated and its three steps:

- Creating path plan
- Assignment of point on path which will keep the motion on path at the process of observing
- Speed and angular velocity generation (by means of robot's kinematics)
- Conversion of speed and angular velocity to rotational speed of wheels.

A player must be able to convert the component of straight line speed and angular velocity to desired rotation speed of individual wheels. It is assumed during the process of computing that the player drives on horizontal mat and wheels are in constant contact with surface. Wheel speed which is required for output data to control part o speed are computed as follows:

$$v_L = v - \omega \cdot \frac{a}{2} \quad v_R = v + \omega \cdot \frac{a}{2} \tag{7.5}$$

A robotic player disposes of two wheels which are connected to two independent DC motors by gearbox. Control input for the motors is power PWM signal from control microcontroller. PSD controller is used to achieve the right motor speed. Required speed (computed form formulas) and real speed of motors (processed output of encoder) are inputs to regulator (Figs. 7.11 and 7.12).

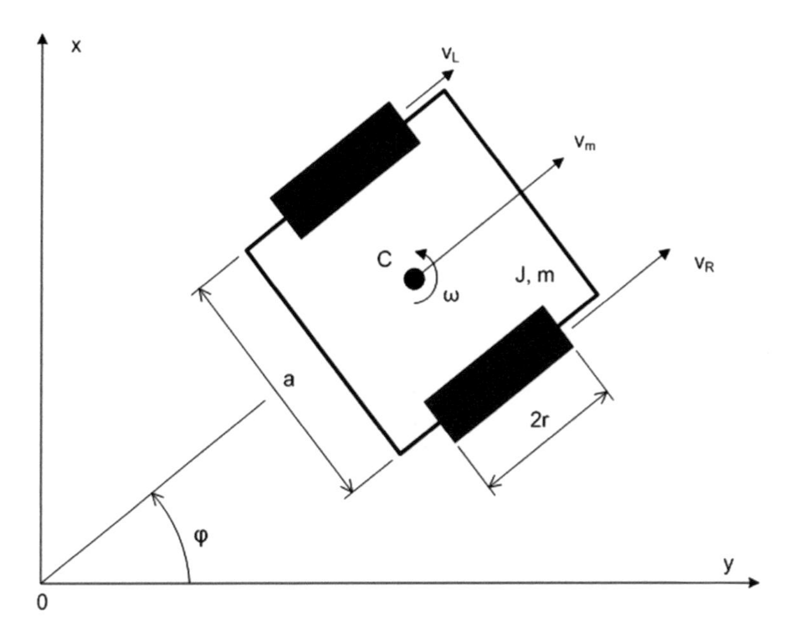

Fig. 7.11 Robot 75 mm × 75 mm for Mirosot category

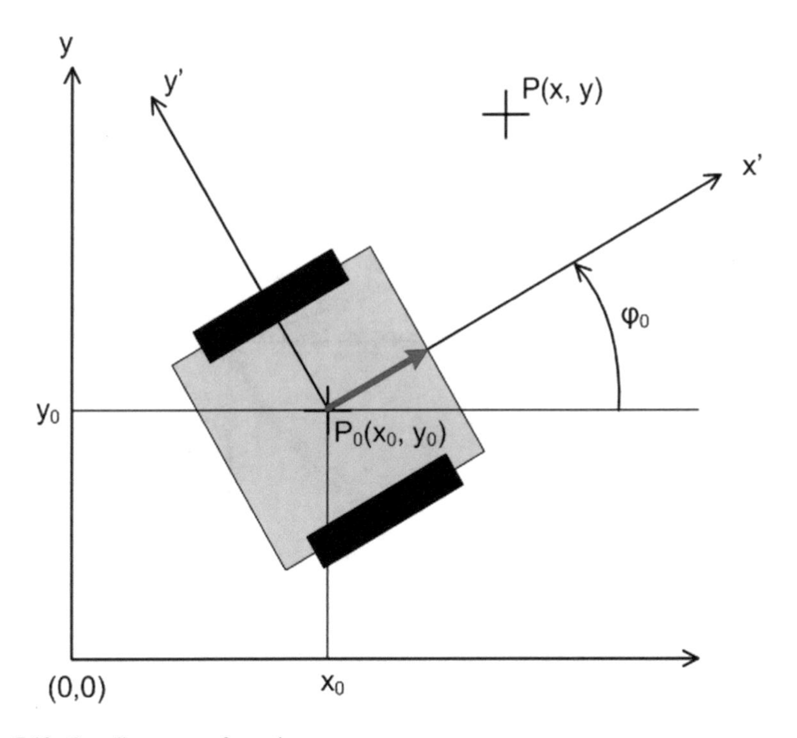

Fig. 7.12 Coordinates transformation

7.10.1 Players' Movement Control Following the Given Trajectory

Mobile robots used for robotic soccer are limited by the size of cube 75 mm. Kinematics is derived from this limitation. As one of the most advanced way of movement appeared to be two-wheeled chassis even if the experiments with tracked robot and four-wheeled robot have been made in the beginning. Two-wheeled gear is undemanding regarding switching motor rotation to the wheel. Robot with two-wheeled chassis is quite adroit and is capable of turning around on one place without slipping at zero translational speed. Problems appear at higher speed and accelerations. Not very convenient mass distribution and thus inconvenient location of the centre may be made disadvantageous by the robots comparing to the others. Robot's motion following circle trajectory with radius ρ (Fig. 7.13) is gained from the rate of speed v and angular velocity ω:

$$\rho = \frac{v}{\omega} \tag{7.6}$$

During the process of path planning, it is possible to compute the next position of a robot for the curve in the following procedure. Position of the robot in time 0 je $P_0 = [x_0, y_0, \varphi_0]$. In case the robot moves at speed v and angular velocity ω, the

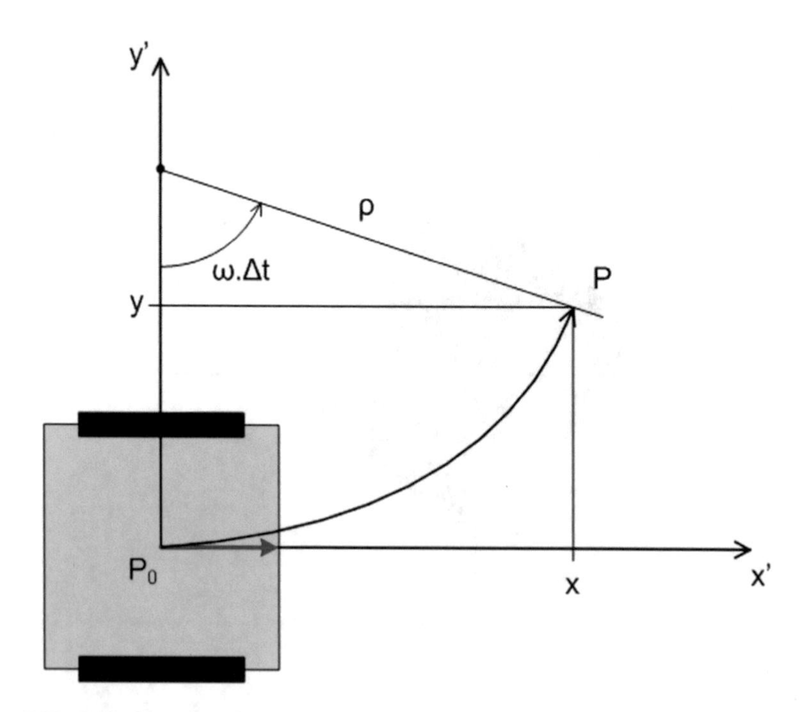

Fig. 7.13 Orbicular path trajectory

next position shall be P = f(P_0, v, ω, Δt). The trajectory between the start point and the final point shall have the shape of circular curve with radius ρ.

In order to simplify computations, new coordinate system has been introduced (Fig. 7.12). Its zero point is situated in the centre of a robot. New coordinate system is shifted in the direction of axis x by x_0 and direction of axis y by y_0. The slew of coordinate system is performed in a way to make sure that the axis x is tangent to the current path and thus coordinates of new coordinate system are:

$$\begin{aligned} x' &= x \cdot \cos\varphi_0 + y \cdot \sin\varphi_0 + x_0 \cdot \cos\varphi_0 - y_0 \cdot \sin\varphi_0 \\ y' &= -x \cdot \sin\varphi_0 + y \cdot \cos\varphi_0 + x_0 \cdot \sin\varphi_0 - y_0 \cdot \cos\varphi_0 \end{aligned} \tag{7.7}$$

Inverse transformation:

$$\begin{aligned} x &= x' \cdot \cos\varphi_0 - y' \cdot \sin\varphi_0 + x_0 \\ y &= y' \cdot \sin\varphi_0 + y' \cdot \cos\varphi_0 + y_0 \end{aligned} \tag{7.8}$$

New position after the time interval Δt:

$$\begin{aligned} x' &= \rho \cdot \sin(\omega \cdot \Delta t) \\ y' &= \rho(1 - \cos(\omega \cdot \Delta t)) \end{aligned} \tag{7.9}$$

Inverse transformation application:

$$\begin{aligned} x &= \frac{v}{\omega} \cdot (\sin(\omega \cdot \Delta t)\cos\varphi_0 - (1 - \cos(\omega \cdot \Delta t)\sin\varphi_0)) + x_0 \\ y &= \frac{v}{\omega} \cdot (\sin(\omega \cdot \Delta t)\sin\varphi_0 + (1 - \cos(\omega \cdot \Delta t)\cos\varphi_0)) + y_0 \\ \varphi &= \omega \cdot \Delta t + \varphi_0 \end{aligned} \tag{7.10}$$

For ω → 0 new position:

$$\begin{aligned} x' &= v \cdot \Delta t \\ y' &= 0 \end{aligned} \tag{7.11}$$

Application on inverse transformation:

$$\begin{aligned} x &= v \cdot \Delta t \cdot \cos\varphi_0 + x_0 \\ y &= v \cdot \Delta t \cdot \sin\varphi_0 + y_0 \\ \varphi &= \varphi_0 \end{aligned} \tag{7.12}$$

7.10.2 Navigation of a Player to the Point Without Resultant Trajectory

At first the robot is swiveled towards the target point by means of change of angular velocity. In case the angular discrepancy is smaller than threshold angle the robot starts accelerating:

$$\begin{aligned} \Delta\varphi &= \varphi_T - \varphi_R \\ \omega &= \frac{K_P}{\Delta T} \cdot \Delta\varphi \end{aligned} \tag{7.13}$$

where ΔT is reading time and Kp represents reinforcement of regulator [16].

Speed increases up to the selected maximum speed. If the angular discrepancy is higher than limit angle, speed is decreasing (Fig. 7.14).

7.10.3 Navigation of a Player Towards the Point

The motion trajectory for player's navigation towards the point is depicted in the Fig. 7.15.

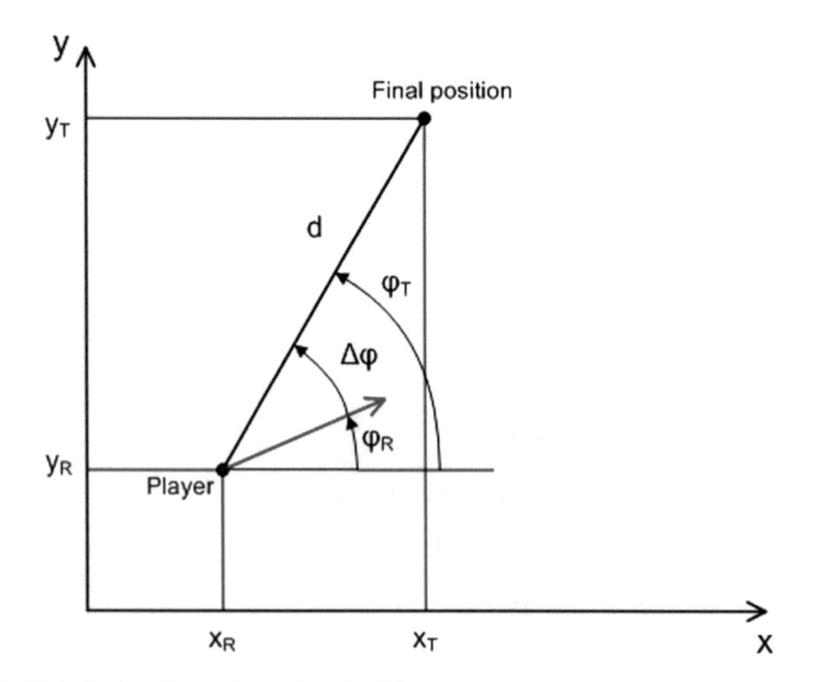

Fig. 7.14 Final position without orientation [18]

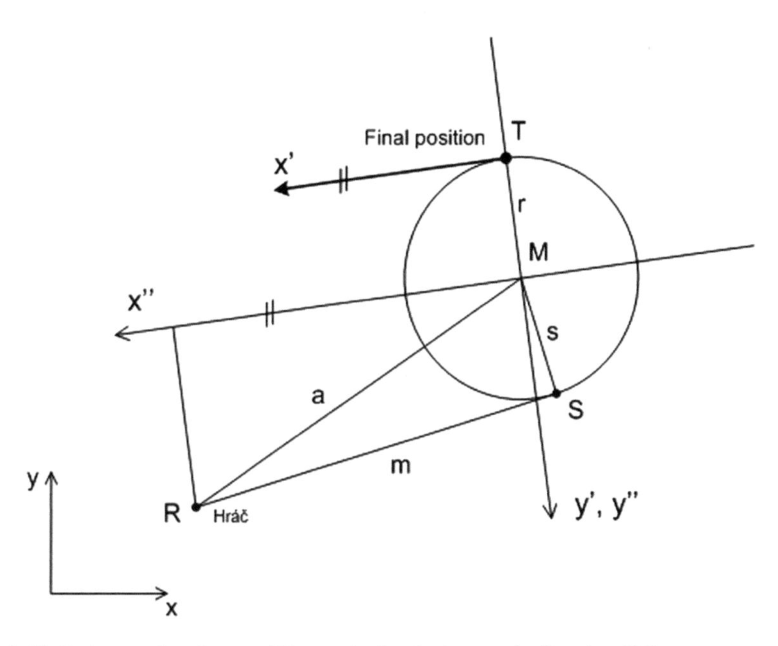

Fig. 7.15 Trajectory for player guiding to final point by certain direction [18]

Trajectory of the player's movement is composed of the movement following line and movement following circle to reach final position of a player with specific orientation. Reached point S has the coordinates [18] (Fig. 7.16):

$$s_{y''} = \frac{a_{y''} \cdot r^2 + \sqrt{a_{x''} \cdot r^2 + a_{x''}^2 \cdot a_{y''}^2 \cdot r^2 - a_{x''}^2 \cdot r^4}}{a_{x''}^2 + a_{y''}^2}$$
$$s_{x''} = \frac{r^2}{a_{x''}} - \frac{s_{y''} \cdot a_{y''}}{a_{x''}} \tag{7.14}$$

φ_R current direction
φ_T orientation in target point
φ_M vector angle towards the centre of circle
φ_S tangent angle in the initial point
φ_1 secant angle
φ_2 angle between secant and radius
$\Delta\varphi$ angle between current direction and tangent in the initial point.

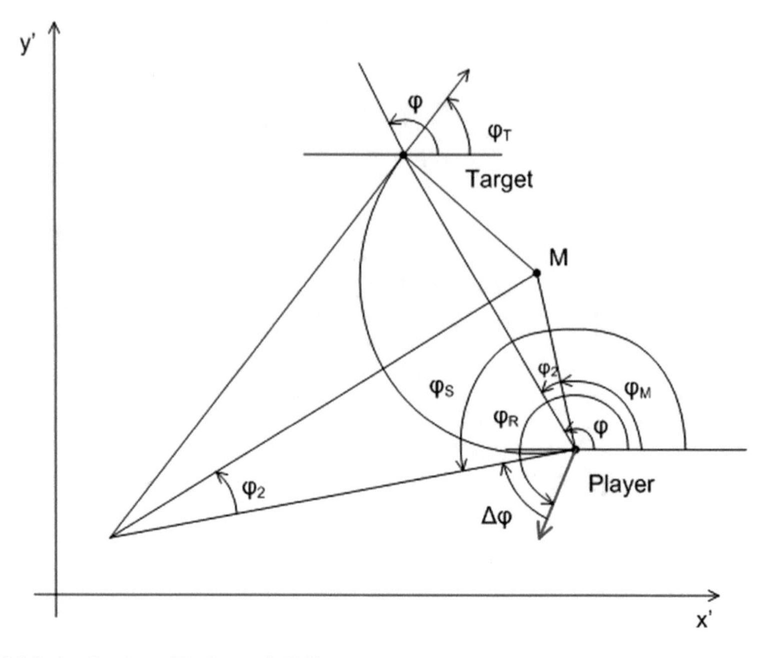

Fig. 7.16 Angles for orbicular path [18]

$$
\begin{aligned}
\varphi_1 &= \arctan\left(\frac{y_T - y_R}{x_T - x_R}\right) \\
\varphi_M &= \arctan\left(\frac{y_M - y_R}{x_M - x_R}\right) \\
\varphi_2 &= \varphi_1 - \varphi_M \\
\varphi_S &= \frac{\pi}{2} + \varphi_M \\
\Delta\varphi &= \varphi_S - \varphi_R
\end{aligned}
\tag{7.15}
$$

Chapter 8
The Methodology of Creating Overall Strategy for Well-Developed MAS

In the stage of strategy s_high$_n$ development it was necessary to define the way the system has to behave and which agents from AH set are to be used. The behavior of the system is defined by the threshold parameters setting for switching and hystereses which determine the transitions between strategic actions s_mid and set of used elementary agents for strategy being formed. After the threshold parameters have been set, the new set of elementary agents was created:

$$\forall s\_high_n \exists AH_n \subset AH;\ |AH_n| \geq 5 \tag{8.1}$$

n…strategy number, n = 1, 2,…the number of created strategies.

Next, elementary agents are to be chosen which belong to each strategic action:

$$\forall s\_mid_{nm} \exists AH_{nm} \subset AH_n;\ |AH_{nm}| = 5 \tag{8.2}$$

n…strategy number

m…strategic action number, m = 1, 2, 3, 4, 5 (absolute defense, defense, midfield play, attack, absolute attack).

$$AH_{nm} = \{a_1, a_2, a_3, a_4, a_5\} \tag{8.3}$$

a_1… agent with the highest priority

⋮

a_5… agent with the lowest priority

One robot belongs to each elementary agent in use:

$$\exists! ID;\ a = ID,\ a \in AH_{nm} \tag{8.4}$$

M. Hajduk et al., *Cognitive Multi-agent Systems*, Studies in Systems, Decision and Control 138, https://doi.org/10.1007/978-3-319-93687-1_8

There are 5 robots of one team on the field. Robots are given by set or robots R:

$$R = \{ID1, ID2, ID3, ID4, ID5\} \tag{8.5}$$

IDn…robot which belongs to an agent an before the process of assignment: for n = 1, 2, 3, 4, 5

$$\begin{aligned} &before\_assignment : a_n = IDn \\ &after\_assignment : a_n = IDm,\ n = 1,2,3,4,5,\ m = 1,2,3,4,5 \end{aligned} \tag{8.6}$$

Robot to agent assignment is done within the rules which were projected for strategic action s_mid. The rules were projected for each elementary agent an, following general key:

$$\begin{aligned} &a_1 \in R \\ &a_2 \in (R - a_1) \\ &a_3 \in (R - a_1 - a_2) \\ &a_4 \in (R - a_1 - a_2 - a_3) \\ &a_5 \in (R - a_1 - a_2 - a_3 - a_4) \end{aligned} \tag{8.7}$$

All of the above mentioned parameters and rules are used by the agent master as its knowledge database and are used for decisions concerning chosen strategic action and assignment of the functions to robots throughout the game.

8.1 Procedure of Creating Overall Strategy

The procedure of creating the overall strategy based on the assumption of using predefined elementary agents is defined by the steps:

(1) *Identification of strategy type and definition of preliminary behavior for individual types.* Strategy type is determined based on the fact whether it is attack or defense action. It is necessary to define the positions of players on the field for individual stages of a game, e.g. if the team is in front of the opponent's goal and is in the ball's possession, three robots attack and one robot serves as back up near the centre circle, etc.
(2) *Switching and hysteresis setting among strategic actions according to the type of overall strategy.* The definition of the borders of switching among strategic actions may considerably affect the strategic behavior of system in respect of the game. In case the borders are considerably forwarded and the team must attack with three or four robots in some stages the final strategy would look defense-like since the attacking phase is not that frequent.

(3) *Elementary agents set choice and their selection for each strategic action in such a way that just 5 elementary players are assign to just one strategic action.* It is necessary to define precisely which elementary agents must exist in which phase. Thus the strategy is defined in strategic actions following the point 1.
(4) *Priority designation of elementary agents' assignment to the robots for each strategic action.* The priority of elementary agents is very important regarding their accurate assignment to the robots. In case the definition is incorrect, e.g. substitution of the goalkeeper and the defender at the moment the opponent is striking the goal or in case the interchange of attackers occurs at the moment one of them steals the ball, it would lead to the loss and later attack from the opponent's team.
(5) *Modification of general key how to assign elementary agents to the robots on the priority basis given by strategic action.* Transformation of general key on the priority basis in such a way that the agent with the highest priority was on top and the agent with the lowest priority as low as possible. The key takes into account the possibility of agents' interchange for individual robots, i.e. that the change of players' positions is defined in such a manner that the change of two furthest positions is not possible (e.g. goalkeeper and attacker shall never change directly).
(6) *Creating algorithm of rules how to assign in accordance with the key and its implementation to the body of the agent master.* The creation of algoritmical module of the agent master. Its decision-making referring to agents' assignment to robots will be based on the key created in point 5. The module agent master and its decision-making on the selection of strategic action will remain unchanged.

It is possible to create many strategies from which two are mentioned in the following chapter by the stated procedure. It is also possible to modify some of the elementary agents or create new ones which would be implemented in the overall strategy in order to extend the possibilities of application if required.

Chapter 9
Analysis of Strategic Behavior of MAS Depending on the Basic Strategy Adjustment (Examples of Designed and Applied Strategies)

This chapter deals with the detailed description of system's behavior for selected basic strategies chosen by the user. Settings for the individual strategies from the user's point of view are listed at the beginning of the description of each strategy. There are sets of elementary agents described, from which the agent master chooses the elementary agents and the rules of ID assignment to each agent with reference to the sequence of assignment.

9.1 Strategy No. 1—Defense

The user chooses the strategy in the driver by combining the options in the settings of strategies (Fig. 6.3):

- DEFENSE—AGGRESIVITY → active
- DEFENNSE—STYLE → side by side
- BACK UP → active
- ATTACK → one rebounds in the centre

These items shall be selected when looking at the combinations (Fig. 6.4):

- 2 cooperating strikers
- the striker without the ball moves in the centre in front of the opponent's goal
- 2 defenders in front of the defense zone are active in blocking the opponent
- midfielder in the centre of the field is active in moving to the sides

Chosen strategy is one of the most defense-based strategies implemented in the program. During the process of attack, maximum two strikers attack with one midfielder. All threshold values of switching strategic actions are moved as much as possible towards the opponent's defense zone. It is being considered to set the active defense and back-up with selected defenders side by side (right and left) in this strategy.

M. Hajduk et al., *Cognitive Multi-agent Systems*, Studies in Systems, Decision and Control 138, https://doi.org/10.1007/978-3-319-93687-1_9

The agent master disposes of this set of elementary agents:

$$AH_1 = \{b, o, op, ol, z, u, ud, us\} \tag{9.1}$$

The agent master uses 8 elementary agents for the strategy. Their behaviour is known. Finally, the agent decides on their assignment to ID on the field. In the previous chapter it is stated that the agent must decide on one of the actions smid from the set of strategic actions Sa and and assigns an agent to ID from the set AH_n based on the chosen strategy shigh (Fig. 9.1).

9.1.1 Total Defense

The following rules of assigning the robots to elementary agents players from the set, which is the subset of AH_1 and contains just 5 elementary agents belonging to the given action for the chosen basic strategy apply to the total defense strategic action:

- goalkeeper
- defender
- right back
- left back
- defensive midfield

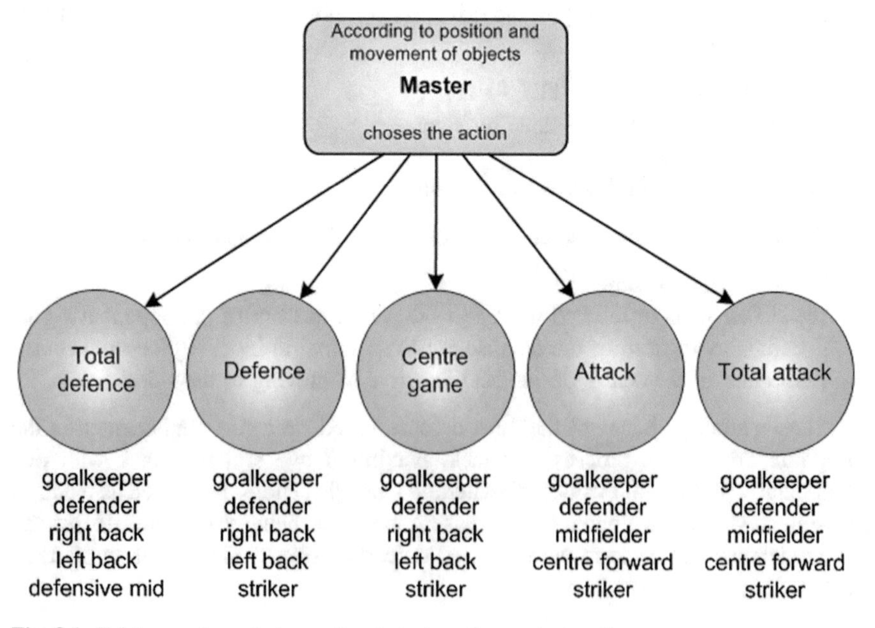

Fig. 9.1 Existence of agent players for strategic actions—strategy 1

The key to assign the robots to the elementary agents:

$$
\begin{aligned}
& b \in \{ID1, ID2\} \\
& b = ID1 \Rightarrow \left\{ \begin{array}{l} o \in \{ID2, ID3, ID4, ID5\} \\ op \in (\{ID2, ID3, ID4, ID5\} - o) \\ ol \in (\{ID2, ID3, ID4, ID5\} - o - op) \\ ud \in (\{ID2, ID3, ID4, ID5\} - o - op - ol) \end{array} \right\} \\
& b = ID2 \Rightarrow \left\{ \begin{array}{l} o = ID1 \\ op \in \{ID3, ID4, ID5\} \\ ol \in (\{ID3, ID4, ID5\} - op) \\ ud \in (\{ID3, ID4, ID5\} - op - ol) \end{array} \right\}
\end{aligned}
\tag{9.2}
$$

Assignment of robots to elementary agent (Fig. 9.2) is performed in the following sequence:

1. goalkeeper
2. defender
3. right back
4. left back
5. defensive midfield

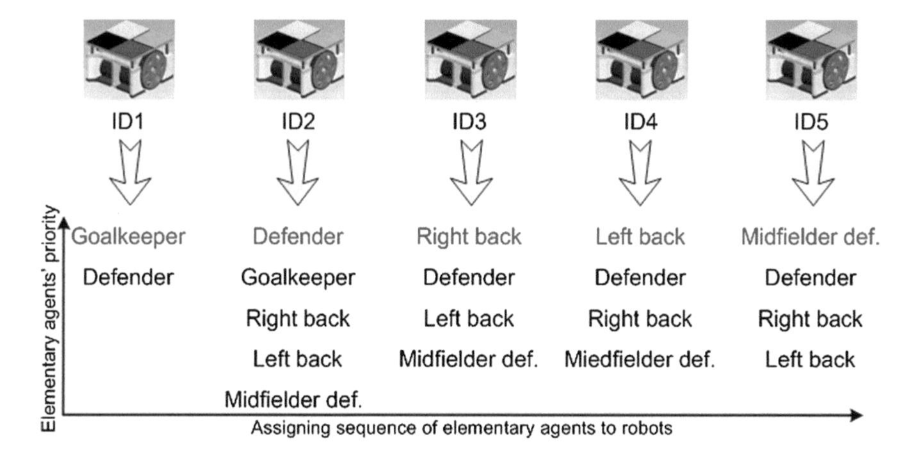

Fig. 9.2 Assigning agent players for total defence action—strategy 1

The rules of players' creation by assigning the robots to the elementary agent players:

$$
\begin{aligned}
& b = IDn_1 \Leftrightarrow \text{Sentence\_of\_position}(IDn_1, b); n_1 = 1,2 \\
& o = IDn_2 \Leftrightarrow \text{Sentence\_of\_position}(IDn_2, o) \wedge o \neq b, n_2 = 1,2,3,4,5 \\
& op = IDn_3 \Leftrightarrow \text{Sentence\_of\_position}(IDn_3, op) \wedge op \neq (b \vee o), n_3 = 2,3,4,5 \\
& ol = IDn_4 \Leftrightarrow \text{Sentence\_of\_position}(IDn_4, ol) \wedge ol \neq (b \vee o \vee op), n_4 = 2,3,4,5 \\
& ud = IDn_5 \Leftrightarrow \text{Sentence\_of\_position}(IDn_5, ud) \wedge ud \neq (b \vee o \vee op \vee ol), n_5 = 2,3,4,5
\end{aligned} \tag{9.3}
$$

Sentence_of_position (x, a) = Robot x has the advanced absolutely and relatively (referring to the ball) positional and speed parameters for the elementary agent a.

At first, the master assigns the goalkeeper to the robot which was the goalkeeper last time (ID1) by the rules and sequence. If its position as the goalkeeper is worse than the position of robot defender at that time, the agent decides on relevant changing of positions of robots. If it is better, then the agent assigns the goalkeeper to the robot (ID2) which was the defender previously. Thus the dynamic interchange of positions among robot players is performed (goalkeeper becomes defender and vice versa). The elementary agent defender may be assigned to any robot, which has more advanced position for the defender in case the defender was not assigned to the robot ID1 yet after the consideration has been taken by the master. In the following sequence the master further assigns agent players to the robots. In the end, the master announces its decisions to all the agents and waits for their arrival of next frame. Simplified algorithm performing the described decisions is depicted in the Figs. 9.3 and 9.4.

9.1.2 Defense

Within the chosen basic strategy the elementary agents from set AH1 are assigned to this action:

- goalkeeper
- defender
- right back
- left back
- striker

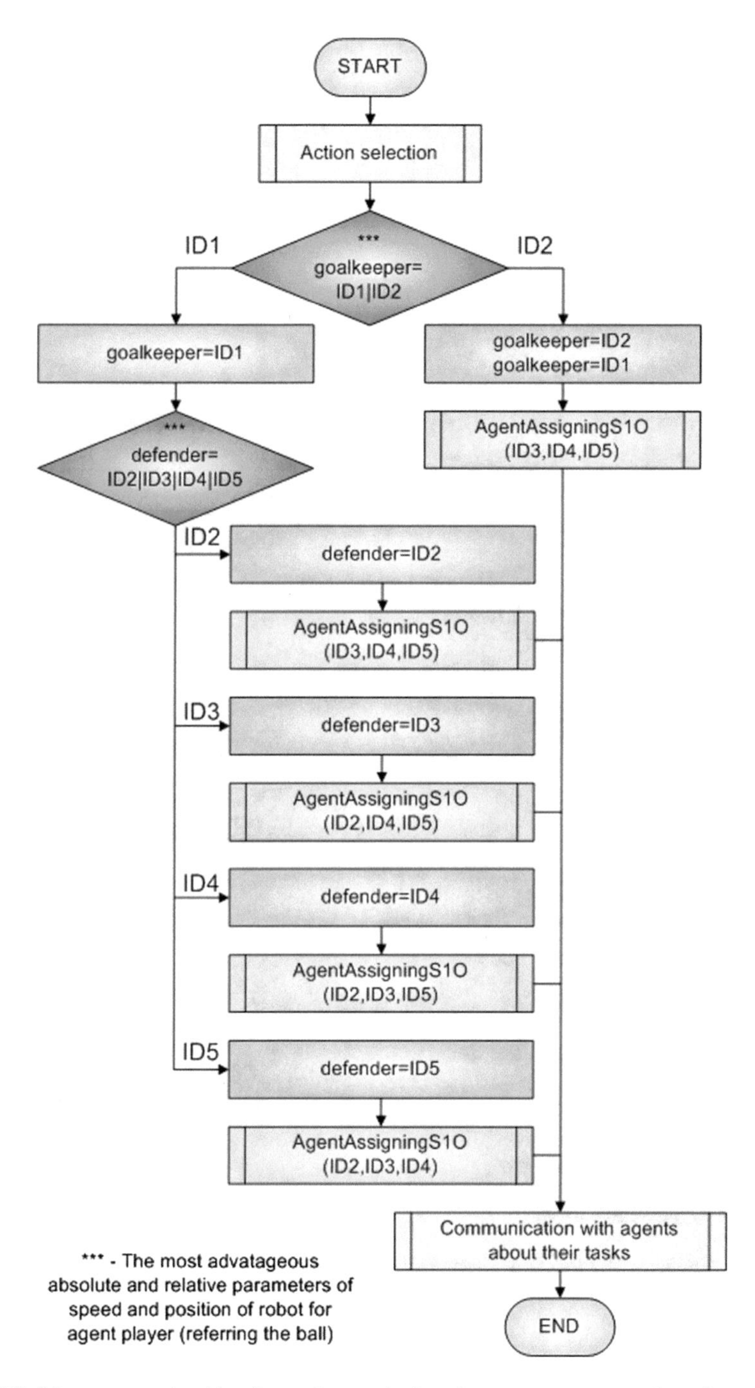

Fig. 9.3 Master agent algorithm focused on assigning elementary agents to robots for strategic action of total defence—strategy 1

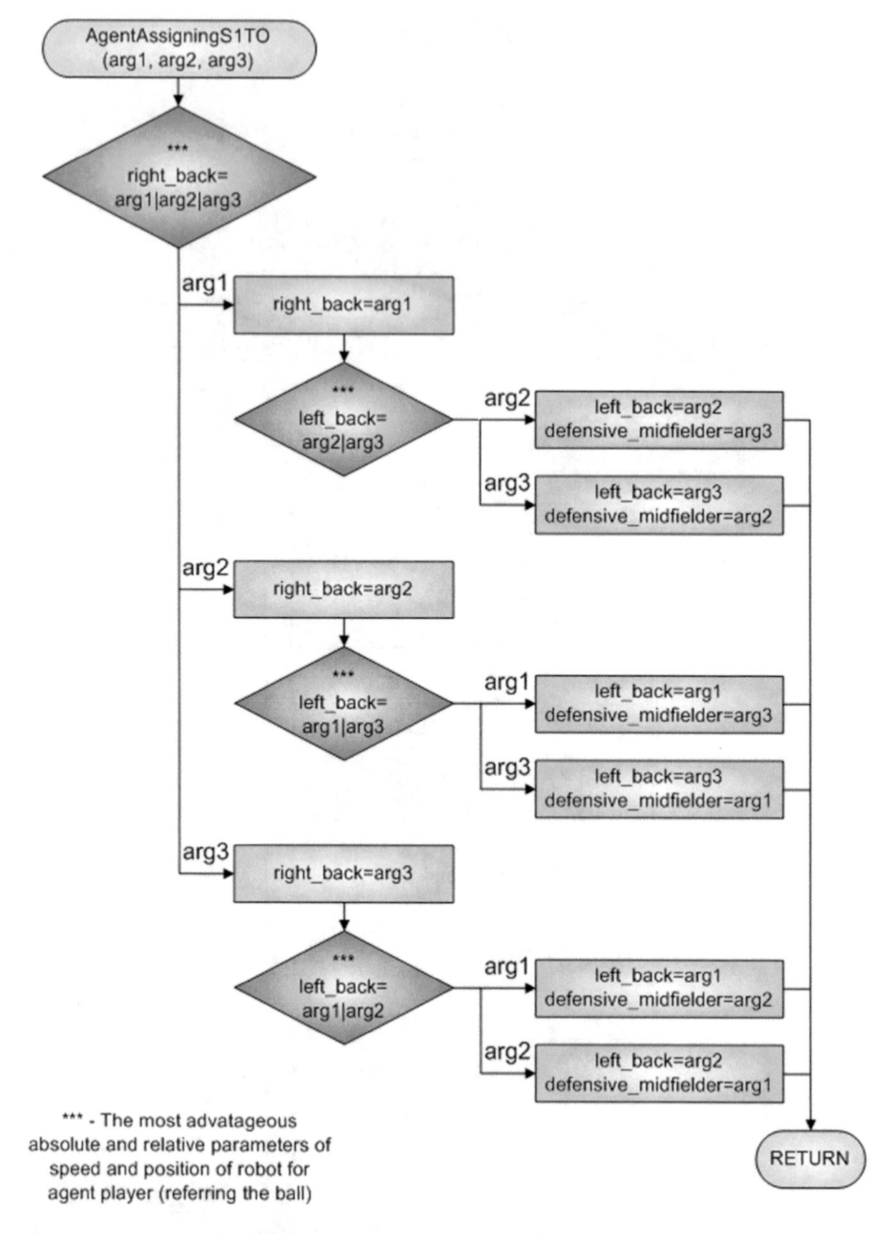

Fig. 9.4 Algorithm of a function "Agent assigning S1TO"—strategy 1

The rules of assigning the robots to the elementary agent players are based on the following principle:

$$
\begin{aligned}
& b \in \{ID1, ID2\} \\
& b = ID1 \Rightarrow \left\{ \begin{array}{l} o \in \{ID2, ID3, ID4, ID5\} \\ op \in (\{ID2, ID3, ID4, ID5\} - o) \\ ol \in (\{ID2, ID3, ID4, ID5\} - o - op) \\ u \in (\{ID2, ID3, ID4, ID5\} - o - op - ol) \end{array} \right\} \\
& b = ID2 \Rightarrow \left\{ \begin{array}{l} o = ID1 \\ op \in \{ID3, ID4, ID5\} \\ ol \in (\{ID3, ID4, ID5\} - op) \\ u \in (\{ID3, ID4, ID5\} - op - ol) \end{array} \right\}
\end{aligned} \tag{9.4}
$$

Assignment of robots to the elementary agents (Fig. 9.5) in the sequence:

1. goalkeeper
2. defender
3. right back
4. left back
5. striker

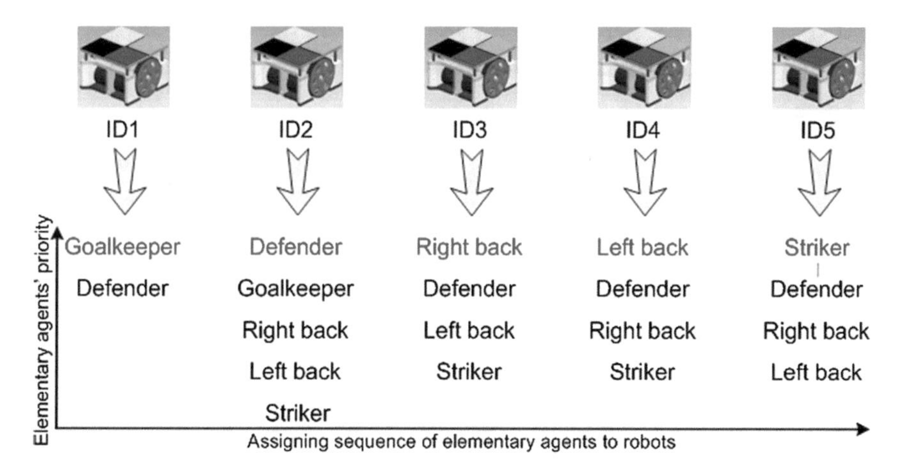

Fig. 9.5 Assigning agent players for defence action—strategy 1

The rules of creation of players by assigning the robots to the elementary agent players:

$$\begin{aligned}
&b = IDn_1 \Leftrightarrow \text{Sentence_of_position}(IDn_1, b); n_1 = 1, 2 \\
&o = IDn_2 \Leftrightarrow \text{Sentence_of_position}(IDn_2, o) \wedge o \neq b, n_2 = 1, 2, 3, 4, 5 \\
&op = IDn_3 \Leftrightarrow \text{Sentence_of_position}(IDn_3, op) \wedge op \neq (b \vee o), n_3 = 2, 3, 4, 5 \\
&ol = IDn_4 \Leftrightarrow \text{Sentence_of_position}(IDn_4, ol) \wedge ol \neq (b \vee o \vee op), n_4 = 2, 3, 4, 5 \\
&u = IDn_5 \Leftrightarrow \text{Sentence_of_position}(IDn_5, u) \wedge u \neq (b \vee o \vee op \vee ol), n_5 = 2, 3, 4, 5
\end{aligned} \tag{9.5}$$

Sentence_of_position (x, a) = robot x has the most advanced absolute and relative positional and speed parameters (with reference to the ball) for the elementary agent *a*.

The assignment of elementary agents to the robots and their sequence for the action defense is the same as it was in case of total defense. The only exception is the change of the elementary agent defensive midfield for strikers. The change takes place because the ball is out of the defense zone. The striker following the ball does not have to enter the defense zone and break the rules related to the high number of defending players in the defense zone or the goal area (Figs. 9.6 and 9.7).

9.1.3 *Midfield Play*

Assigned elementary agents from AH1 are:

- goalkeeper
- defender
- right back
- left back
- striker

The rules of assigning the robots to the elementary agents are based on the following principle (Fig. 9.8):

$$\begin{aligned}
&u \in \{ID5, ID4, ID3\} \\
&ol \in (\{ID5, ID4, ID3\} - u) \\
&op \in (\{ID5, ID4, ID3\} - u - ol) \\
&b \in \{ID1, ID2\} \\
&o \in (\{ID1, ID2\} - b)
\end{aligned} \tag{9.6}$$

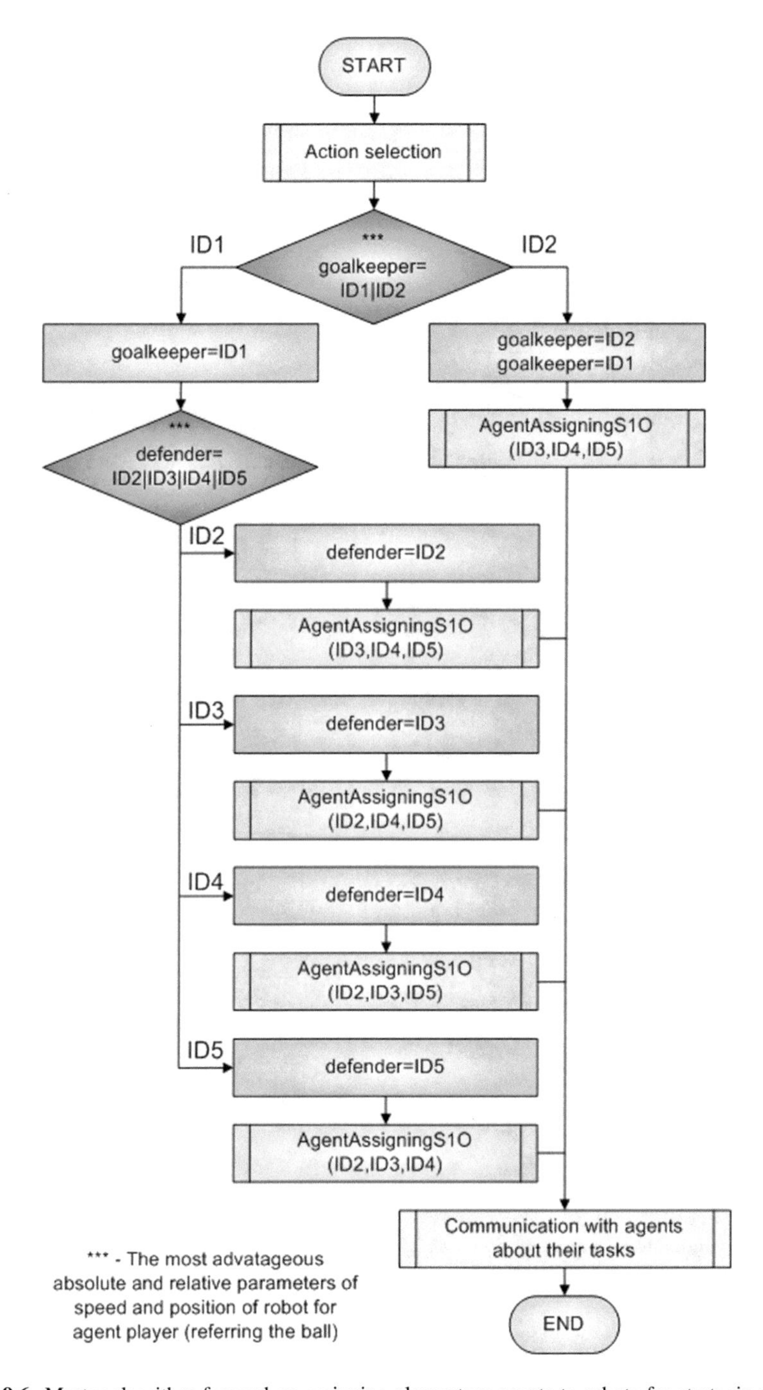

Fig. 9.6 Master algorithm focused on assigning elementary agents to robots for strategic action defence—strategy 1

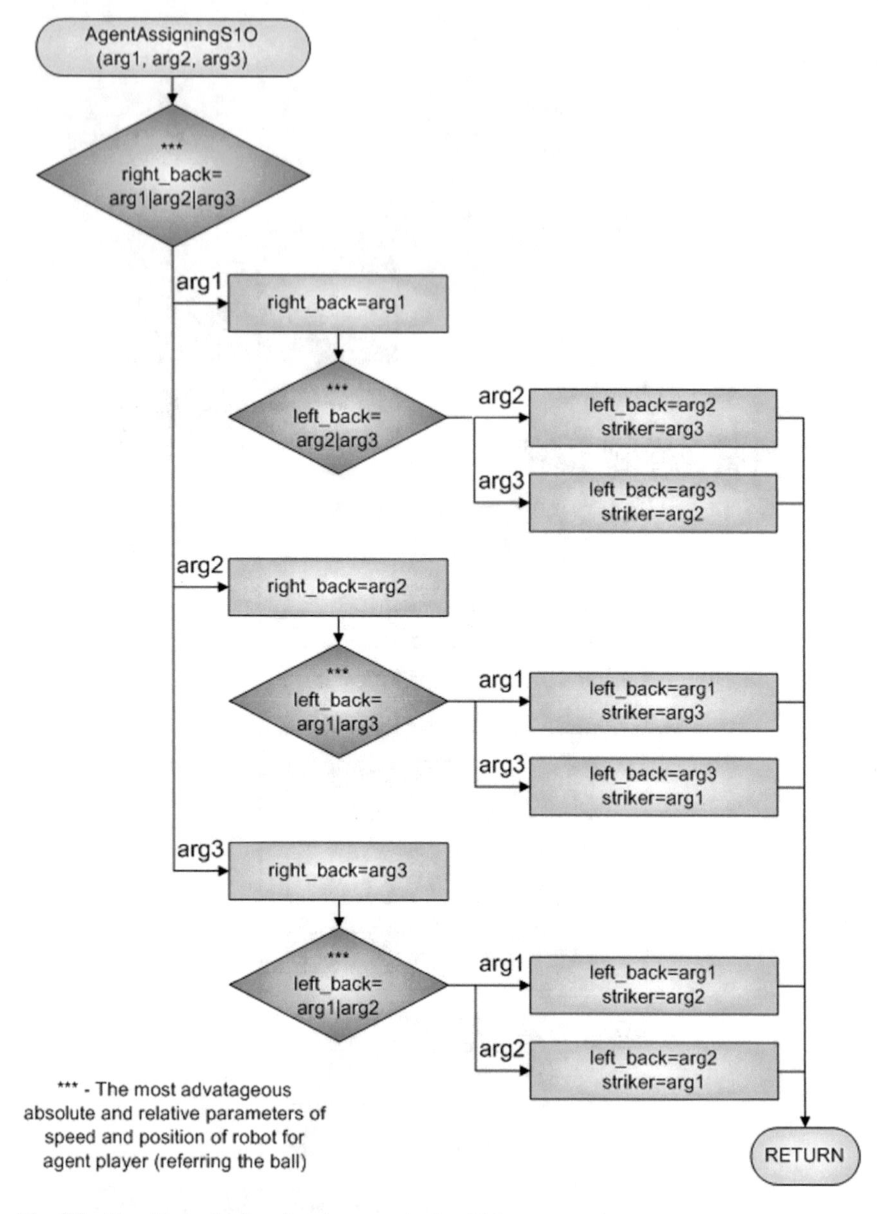

Fig. 9.7 Algorithm of a function Agent assigning S10—strategy 1

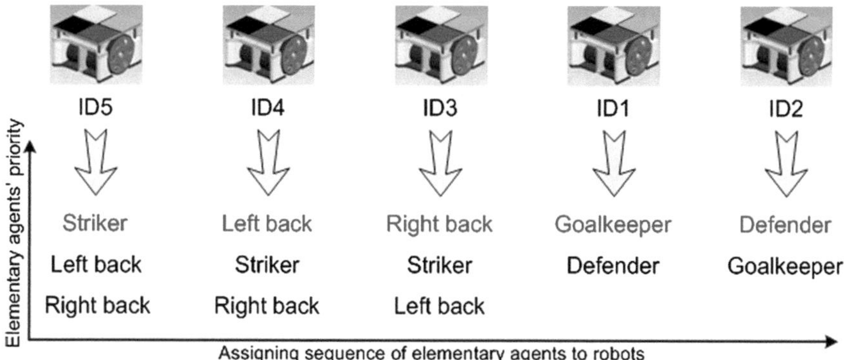

Fig. 9.8 The possibilities of assigning agent players to the action midfield play—strategy 1

The assignment of robots to the elementary agents is performed in the following order:

1. striker
2. left back
3. right back
4. goalkeeper
5. defender

The rules of the creation of players by assigning the robots to the elementary agent players:

$$\begin{aligned}
&u = IDn_1 \Leftrightarrow \text{Sentence_of_position}(IDn_1, u); n_1 = 5, 4, 3\\
&ol = IDn_2 \Leftrightarrow \text{Sentence_of_position}(IDn_2, ol) \wedge ol \neq u, n_2 = 5, 4, 3\\
&op = IDn_3 \Leftrightarrow \text{Sentence_of_position}(IDn_3, op) \wedge op \neq (u \vee ol), n_3 = 5, 4, 3\\
&b = IDn_4 \Leftrightarrow \text{Sentence_of_position}(IDn_4, b), n_4 = 1, 2\\
&o = IDn_5 \Leftrightarrow \text{Sentence_of_position}(IDn_5, o) \wedge o \neq b, n_5 = 1, 2
\end{aligned} \tag{9.7}$$

Sentence_of_position (x, a) = robot x has the most advanced absolute and relative positional and speed parameters for the elementary agent *a*.

The significant change of elementary agents and sequence of robots for assignment might be seen. The master prefers the assignment of the striker to the defenders. The change of sequence of the assignment of defenders is only formal and results from the assumption that ID4 was left back at the transfer from the defense. The goalkeeper and the defender remain assigned to ID1 or ID2. Figure 9.9 shows the algorithm of the agent master for the assignment of the robots to the elementary agent players.

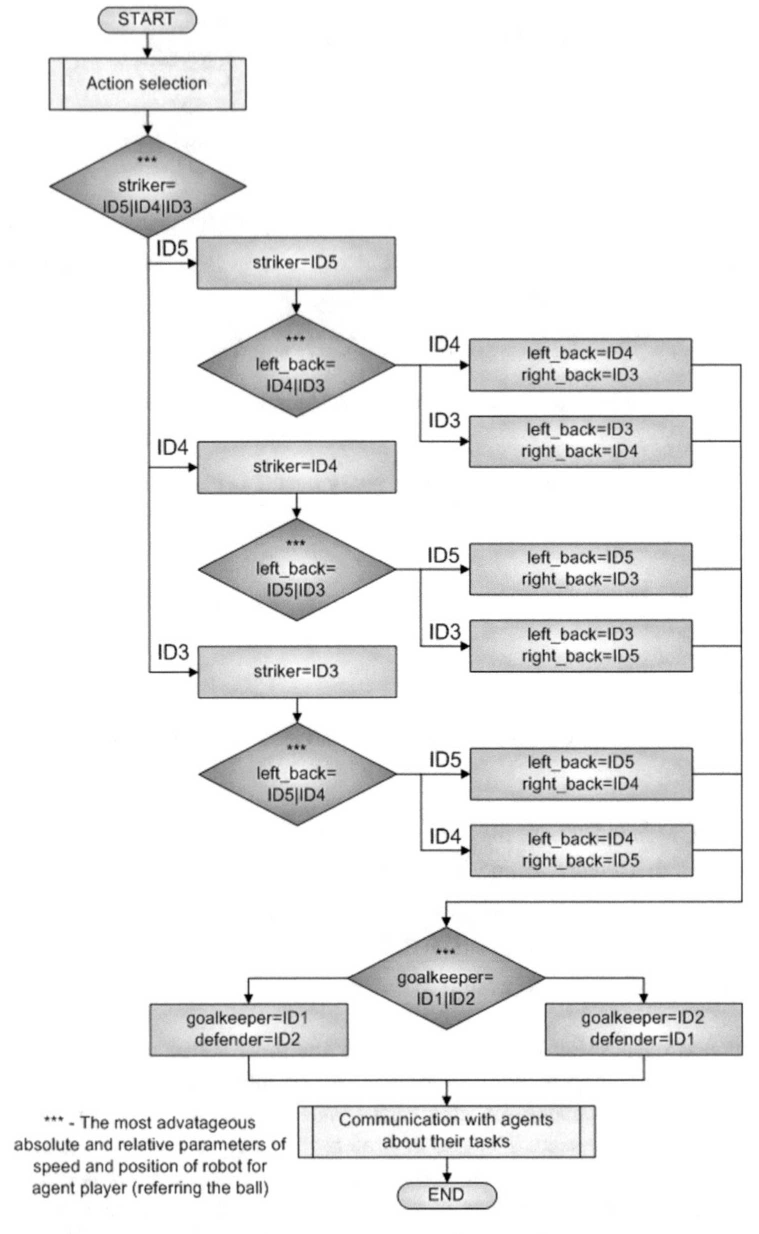

Fig. 9.9 Algorithm of the master focusing on the assignment of elementary agents to the robots for the strategic action midfield play—strategy 1

9.1.4 Attack and Total Attack

Assigned elementary agents from AH1 are:

- goalkeeper
- defender
- midfielder
- centre forward
- striker

The analysis of the strategic behaviour of MAS depending on the basic strategy setting (examples of designed and applied strategies).

The rules of assigning the robots to the elementary agents are based on the following key:

$$\begin{aligned} & u \in \{ID5, ID4, ID3\} \\ & us \in (\{ID5, ID4, ID3\} - u) \\ & z \in (\{ID5, ID4, ID3\} - u - us) \\ & b \in \{ID1, ID2\} \\ & o \in (\{ID1, ID2\} - b) \end{aligned} \tag{9.8}$$

The assigning the robots to the elementary agents (Fig. 9.10) is performed in the following order:

1. striker
2. centre forward
3. midfielder
4. goalkeeper
5. defender

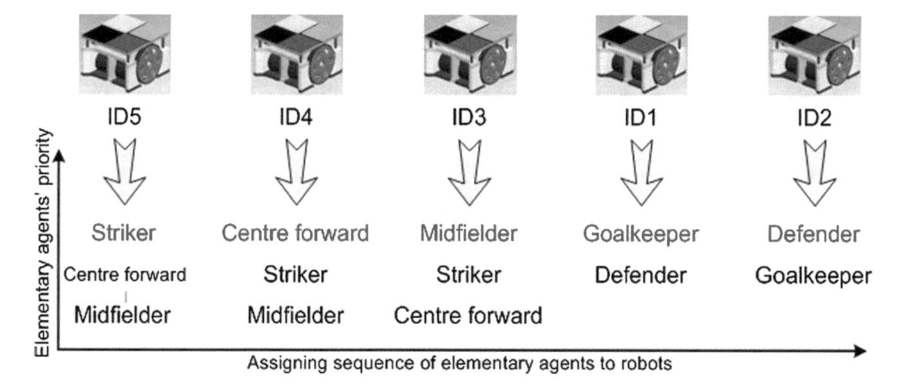

Fig. 9.10 The possibilities of assigning agents players to the action attack and total attack—strategy 1

The rules of the creation of players by assigning the robots to the elementary agent players:

$$\begin{aligned} & u = IDn_1 \Leftrightarrow \text{Sentence\_of\_position}(IDn_1, u); n_1 = 5,4,3 \\ & us = IDn_2 \Leftrightarrow \text{Sentence\_of\_position}(IDn_2, us) \wedge us \neq u, n_2 = 5,4,3 \\ & z = IDn_3 \Leftrightarrow \text{Sentence\_of\_position}(IDn_3, z) \wedge z \neq (u \vee us), n_3 = 5,4,3 \\ & b = IDn_4 \Leftrightarrow \text{Sentence\_of\_position}(IDn_4, b), n_4 = 1,2 \\ & o = IDn_5 \Leftrightarrow \text{Sentence\_of\_position}(IDn_5, o) \wedge o \neq b, n_5 = 1,2 \end{aligned} \quad (9.9)$$

Sentence_of_position (x, a) = robot x has the most advanced absolute and relative positional and speed parameters (referring to the ball) for the elementary agent players.

The Fig. 9.11 shows the part of algorithm of the agent master in which the agent decides on the assignment of the robots to the elementary agent players.

9.2 Strategy No. 2—Offensive Action

The strategy is chosen by the user in the driver by combination of options in the strategy settings (Fig. 6.3):

- DEFFENSE → AGGRESSIVITY
- DEFFENSE → STYLE side by side
- BACK-UP → ACTIVE
- ATTACK → 2 players rebound with centre back

The following items are going to be chosen after checking the combinations in the Fig. 6.4.

- 3 cooperating strikers
- strikers without the ball on the sides in front of the goal
- midfielder in the centre of the field is active when moving to the sides
- 2 defenders in front of the defense zone are standing in front of each other
- Defenders in front of the defense zone are active in blocking the opponent

This strategy is one of two most offense-oriented basic strategies which were implemented to the program. During the attack there are always three strikers with one midfielder. All the threshold values of switching the strategic actions are highly moved towards our defense zone. Within this strategy it is considered to set the active defense and back-up with chosen defenders in front (back defender and centre back).

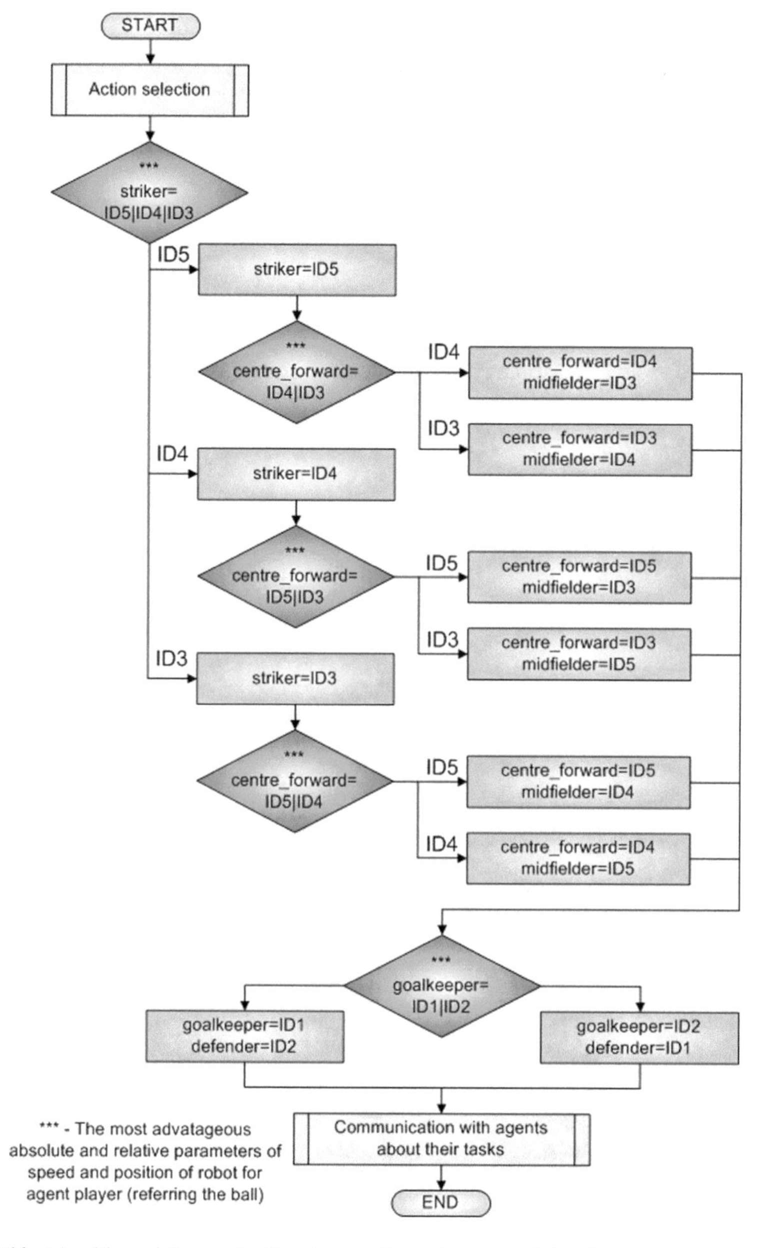

Fig. 9.11 Algorithm of the master focusing on the assignment of the elementary agents to the robots for the strategic action attack and total attack—strategy 1

The agent master disposes of this set of the elementary agents:

$$AH_1 = \{b, o, oz, ov, z, u, ud, up, ul\} \tag{9.10}$$

The agent master uses 9 elementary agents altogether which behaviour is known (Fig. 9.12).

9.2.1 *Total Defense*

Within the chosen basic strategy, the elementary agents assigned to this action from the set AH_2 are:

- Goalkeeper
- Defender
- Back defender
- Centre back
- Defensive midfielder

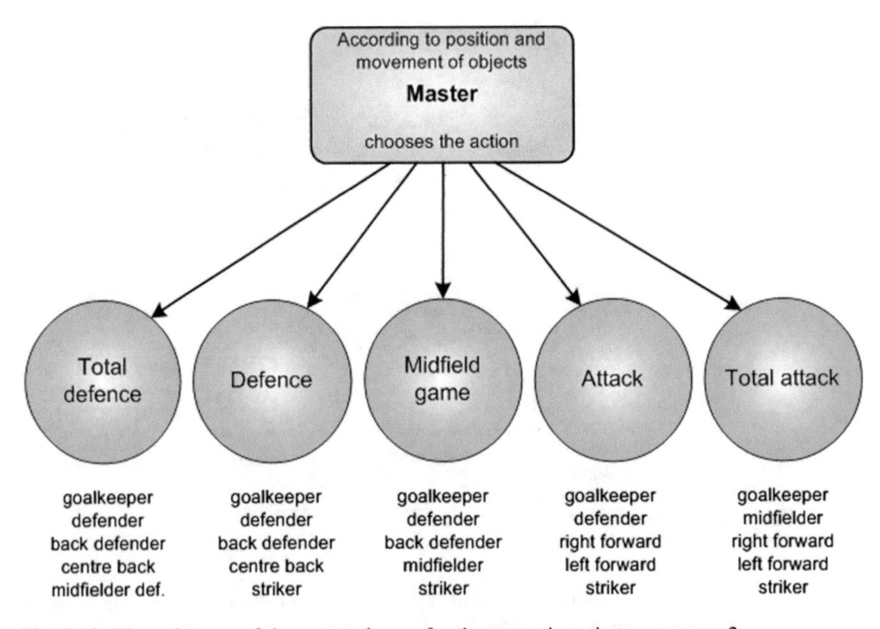

Fig. 9.12 The existence of the agent players for the strategic actions—strategy 2

Te assignment of the robots to the elementary agent players is performed based on the following key:

$$
\begin{aligned}
& b \in \{ID1, ID2\} \\
& b = ID1 \Rightarrow \left\{ \begin{array}{l} o \in \{ID2, ID3, ID4, ID5\} \\ oz \in (\{ID2, ID3, ID4, ID5\} - o) \\ ov \in (\{ID2, ID3, ID4, ID5\} - o - oz) \\ ud \in (\{ID2, ID3, ID4, ID5\} - o - oz - ov) \end{array} \right\} \\
& b = ID2 \Rightarrow \left\{ \begin{array}{l} o = ID1 \\ oz \in \{ID3, ID4, ID5\} \\ ov \in (\{ID3, ID4, ID5\} - oz) \\ ud \in (\{ID3, ID4, ID5\} - oz - ov) \end{array} \right\}
\end{aligned} \tag{9.11}
$$

The assignment of the player to the elementary agents (Fig. 9.13) is performed in the following order:

1. Goalkeeper
2. Defender
3. Back defender
4. Centre back
5. Defensive midfield

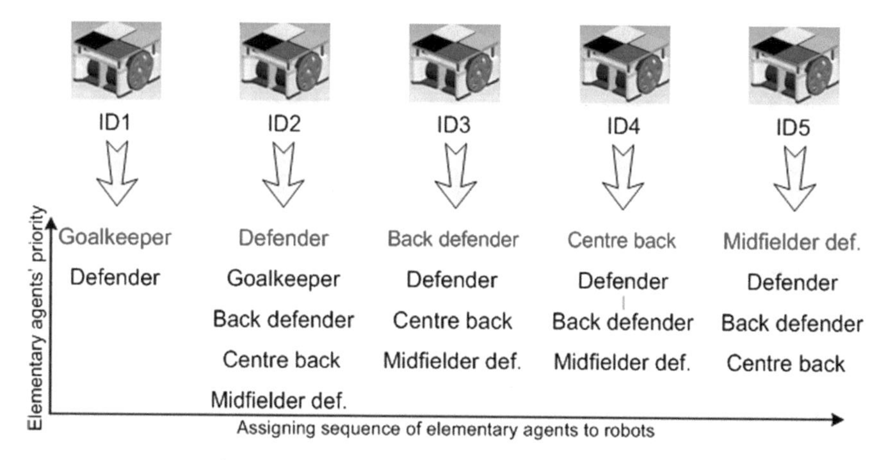

Fig. 9.13 The possibilities of the assignment of the agent players to the action total defense—strategy 2

The rules of the creation of the assignment of the robots to the elementary agent players:

$$
\begin{aligned}
& b = IDn_1 \Leftrightarrow \text{Sentence\_of\_position}(IDn_1, b); n_1 = 1,2 \\
& o = IDn_2 \Leftrightarrow \text{Sentence\_of\_position}(IDn_2, o) \wedge o \neq b, n_2 = 1,2,3,4,5 \\
& oz = IDn_3 \Leftrightarrow \text{Sentence\_of\_position}(IDn_3, oz) \wedge oz \neq (b \vee o), n_3 = 2,3,4,5 \\
& ov = IDn_4 \Leftrightarrow \text{Sentence\_of\_position}(IDn_4, ov) \wedge ov \neq (b \vee o \vee oz), n_4 = 2,3,4,5 \\
& ud = IDn_5 \Leftrightarrow \text{Sentence\_of\_position}(IDn_5, ud) \wedge ud \neq (b \vee o \vee oz \vee ov), n_5 = 2,3,4,5
\end{aligned} \tag{9.12}
$$

Sentence_of_position (x, a) = robot x has the most advanced absolute and relative (referring to the ball) positional and speed parameters for the elementary agent *a*.

The master assigns the goalkeeper to the robot first, which was the goalkeeper previously (ID1) following the rules and sequence (as it was in the strategy no 1). In case its position of goalkeeper is worse at the moment than the position of the robot defender, the agent decides on changing their positions. If it is better, the robot (ID2) is assigned to the goalkeeper which was the defender previously. The elementary agent defender may be assigned (after the consideration has been done by the master) to any robot which has more advanced position for the defender in case the defender has not been assigned to the robot ID1 yet. The master further assigns the other elementary agents to the robots in the mentioned order. In the end, the master announces its decisions to all the agents and waits until the next frame. Simplified algorithm performing the abovementioned decisions is shown in the Figs. 9.14 and 9.15.

9.2.2 *Defense*

Within the chosen basic strategy the following elementary agents are assigned to this action from the set AH_2:

- Goalkeeper
- Defender
- Back defender
- Centre back
- Striker

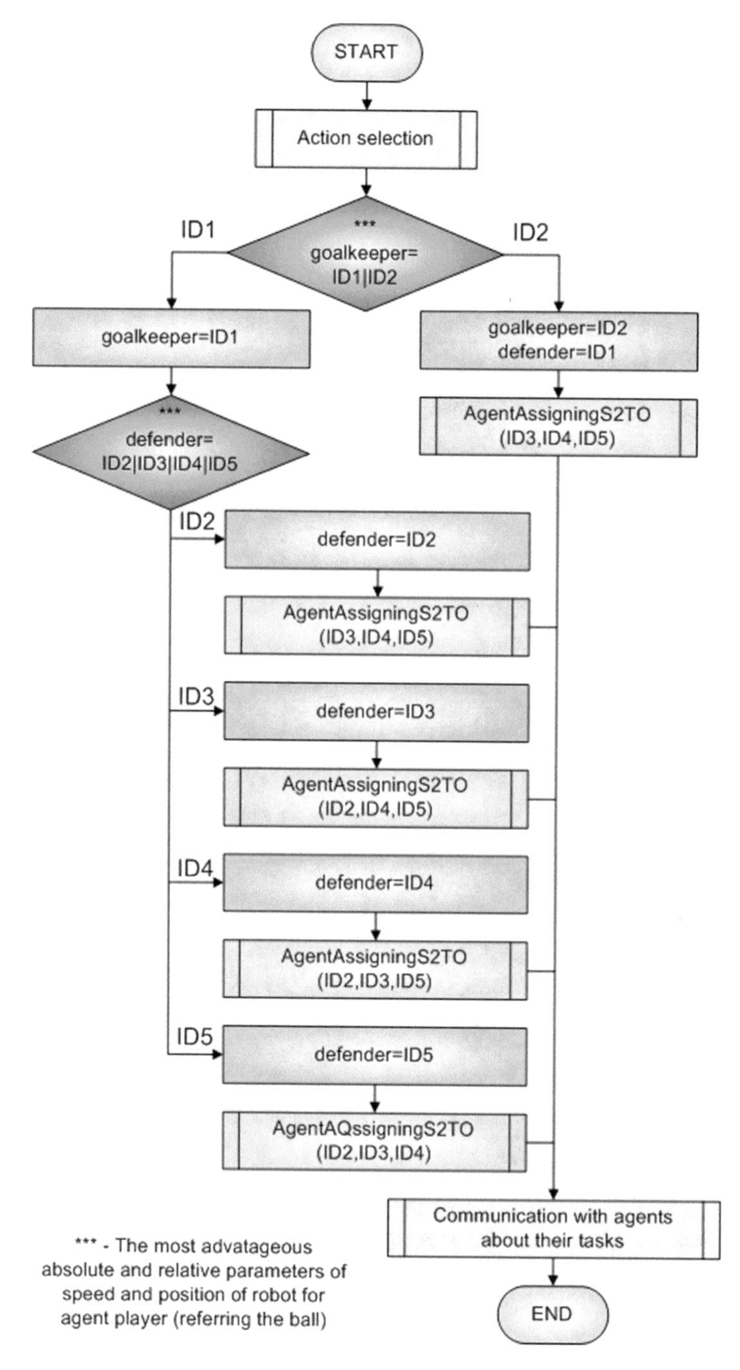

Fig. 9.14 The algorithm of the agent master focusing on the assignment of the elementary agents to the robots for the strategic action total defense—strategy 2

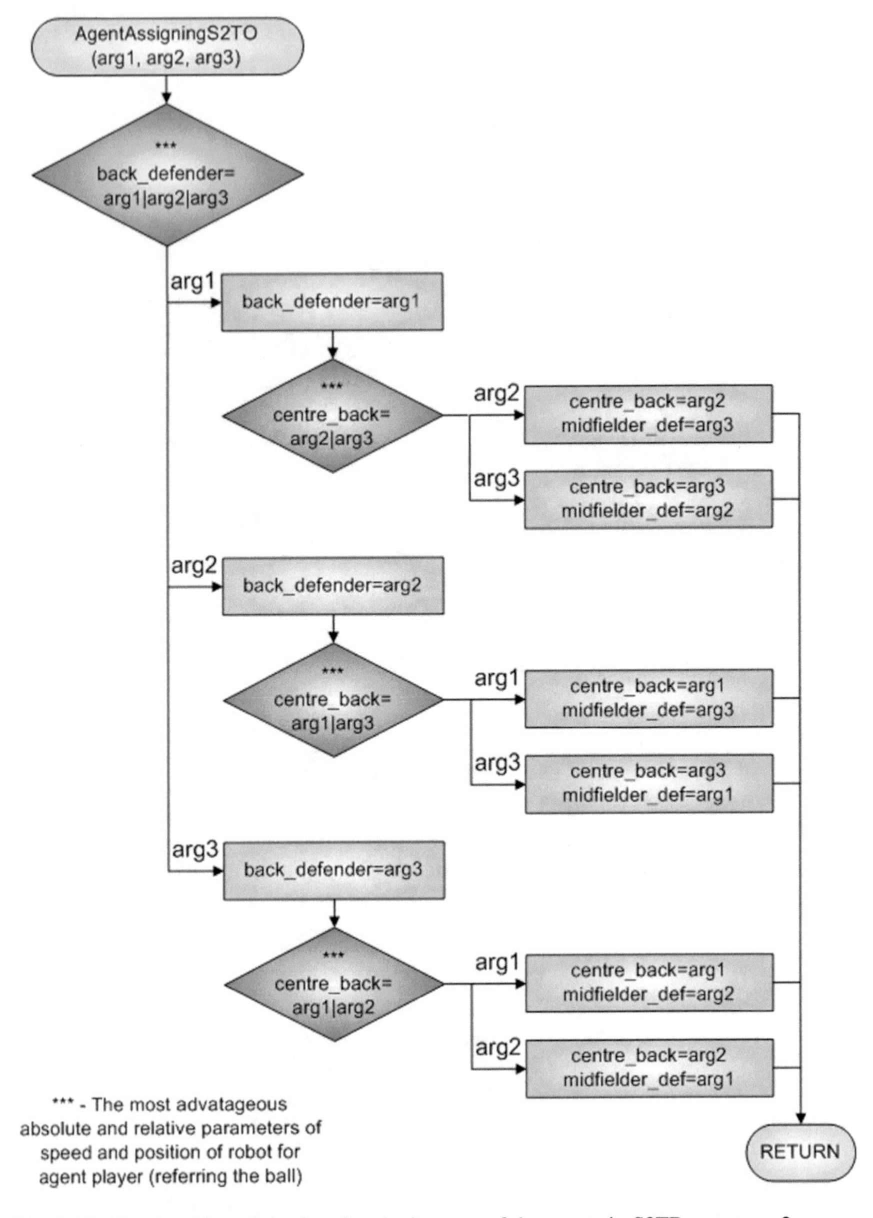

Fig. 9.15 The algorithm of the function Assignment of the agents in S2TD—strategy 2

The rules of the assignment of the robots to the elementary agents following the key:

$$
\begin{aligned}
& b \in \{ID1, ID2\} \\
& b = ID1 \Rightarrow \left\{ \begin{array}{l} o \in \{ID2, ID3, ID4, ID5\} \\ oz \in (\{ID2, ID3, ID4, ID5\} - o) \\ ov \in (\{ID2, ID3, ID4, ID5\} - o - oz) \\ u \in (\{ID2, ID3, ID4, ID5\} - ov - oz - ov) \end{array} \right\} \\
& b = ID2 \Rightarrow \left\{ \begin{array}{l} o = ID1 \\ oz \in \{ID3, ID4, ID5\} \\ ov \in (\{ID3, ID4, ID5\} - oz) \\ u \in (\{ID3, ID4, ID5\} - oz - ov) \end{array} \right\}
\end{aligned}
\tag{9.13}
$$

The assignment of the robots to the elementary agents (Fig. 9.16) is performed in the following order:

1. Goalkeeper
2. Defender
3. Back defender
4. Centre back
5. Striker

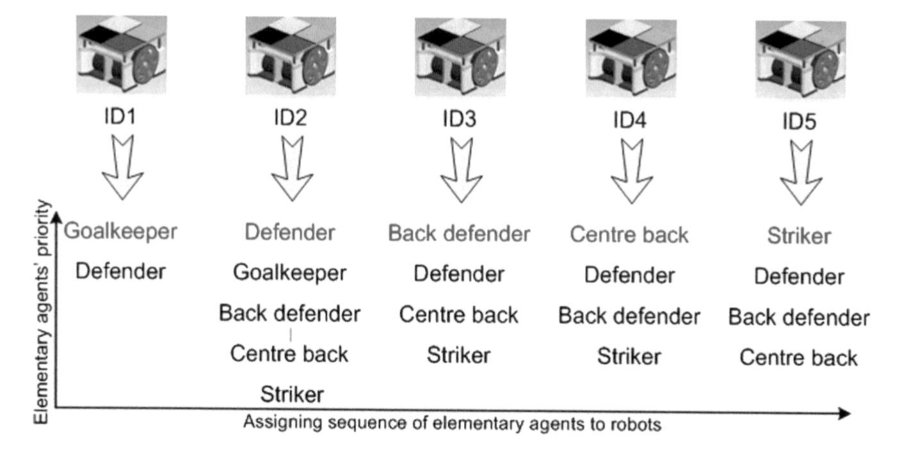

Fig. 9.16 The possibilities of the assignment of the agent players to the action defense—strategy 2

The rules of the creation of players by assigning the robots to the elementary agent players:

$$
\begin{aligned}
& b = IDn_1 \Leftrightarrow \text{Sentence\_of\_position}(IDn_1, b); n_1 = 1, 2 \\
& o = IDn_2 \Leftrightarrow \text{Sentence\_of\_position}(IDn_2, o) \wedge o \neq b, n_2 = 1, 2, 3, 4, 5 \\
& oz = IDn_3 \Leftrightarrow \text{Sentence\_of\_position}(IDn_3, oz) \wedge oz \neq (b \vee o), n_3 = 2, 3, 4, 5 \\
& ov = IDn_4 \Leftrightarrow \text{Sentence\_of\_position}(IDn_4, ov) \wedge ov \neq (b \vee o \vee oz), n_4 = 2, 3, 4, 5 \\
& u = IDn_5 \Leftrightarrow \text{Sentence\_of\_position}(IDn_5, u) \wedge u \neq (b \vee o \vee oz \vee ov), n_5 = 2, 3, 4, 5
\end{aligned}
\tag{9.14}
$$

Sentence_of_position (x, a) = robot x has the most advanced absolute and relative (referring to the ball) positional and speed parameters for the elementary agent *a*.

The assignment of the elementary agents to the robots and their sequence for the action defense is the same as it was in case of total defense (as the strategy No. 1). The only exception is the change of the elementary agent defensive midfield and the striker. The change occurs because the ball is out of the defense zone already. The striker following the ball does not have to enter the defense zone and thus break the rules regarding the high number of defending players in the defense zone or goal area. The part of the agent master and its algorithm with the decision-making process of the assignment of the robots to the elementary agents is depicted in the Figs. 9.17 and 9.18.

9.2.3 The Midfield Play

The assigned elementary agents from AH_2 are:

- Goalkeeper
- Defender
- Back defender
- Midfielder
- Striker

The rules of the assignment of the robots to the elementary agents following the key:

$$
\begin{aligned}
& u \in \{ID5, ID4, ID3, ID2\} \\
& z \in (\{ID5, ID4, ID3, ID2\} - u) \\
& oz \in (\{ID5, ID4, ID3, ID2\} - u - z) \\
& u \neq ID2 \wedge z \neq ID2 \wedge oz \neq ID2 \Rightarrow b \in \{ID1, ID2\}; o \in \{ID1, ID2\} - b \\
& u = ID2 \vee z = ID2 \vee oz = ID2 \Rightarrow b = ID1; o \in (\{ID5, ID4, ID3, ID2\} - u - z - oz)
\end{aligned}
\tag{9.15}
$$

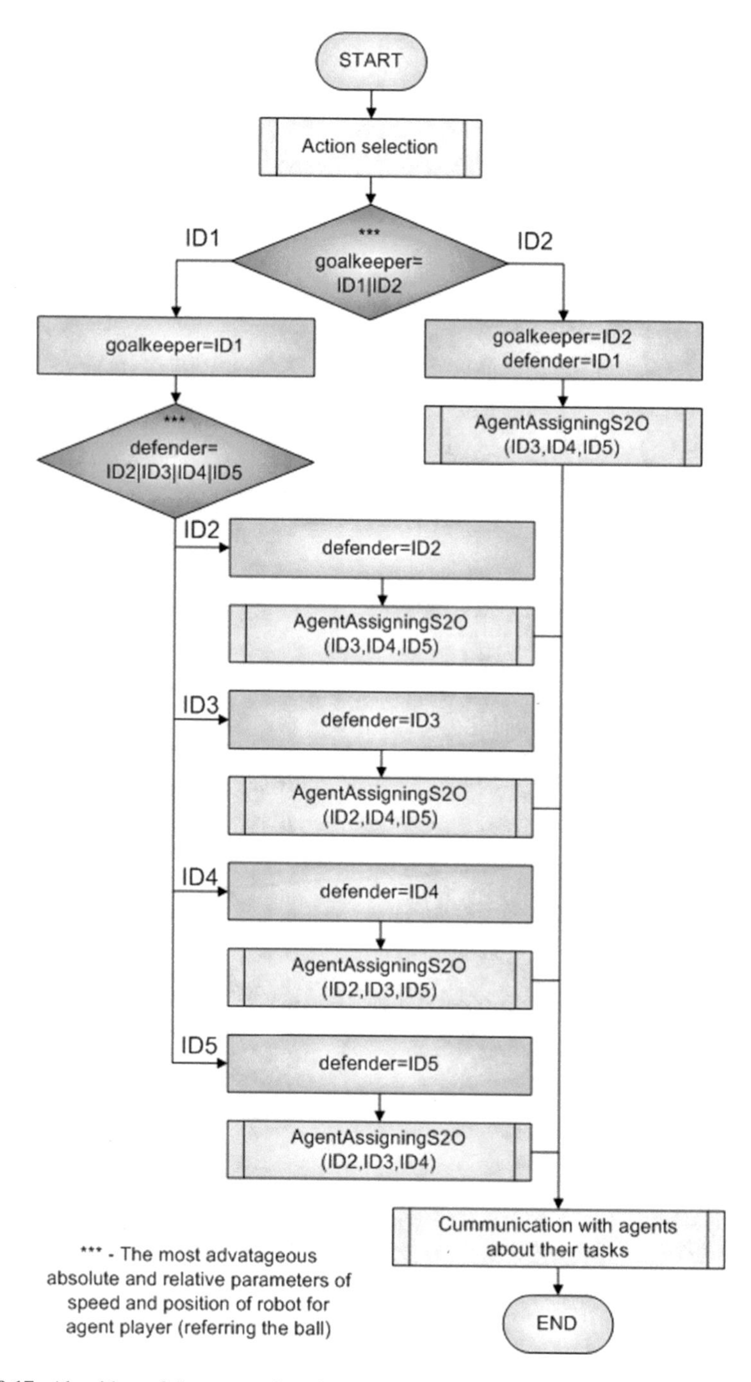

Fig. 9.17 Algorithm of the master focusing on the assignment of the elementary agents to the robots for the strategic action defense—strategy 2

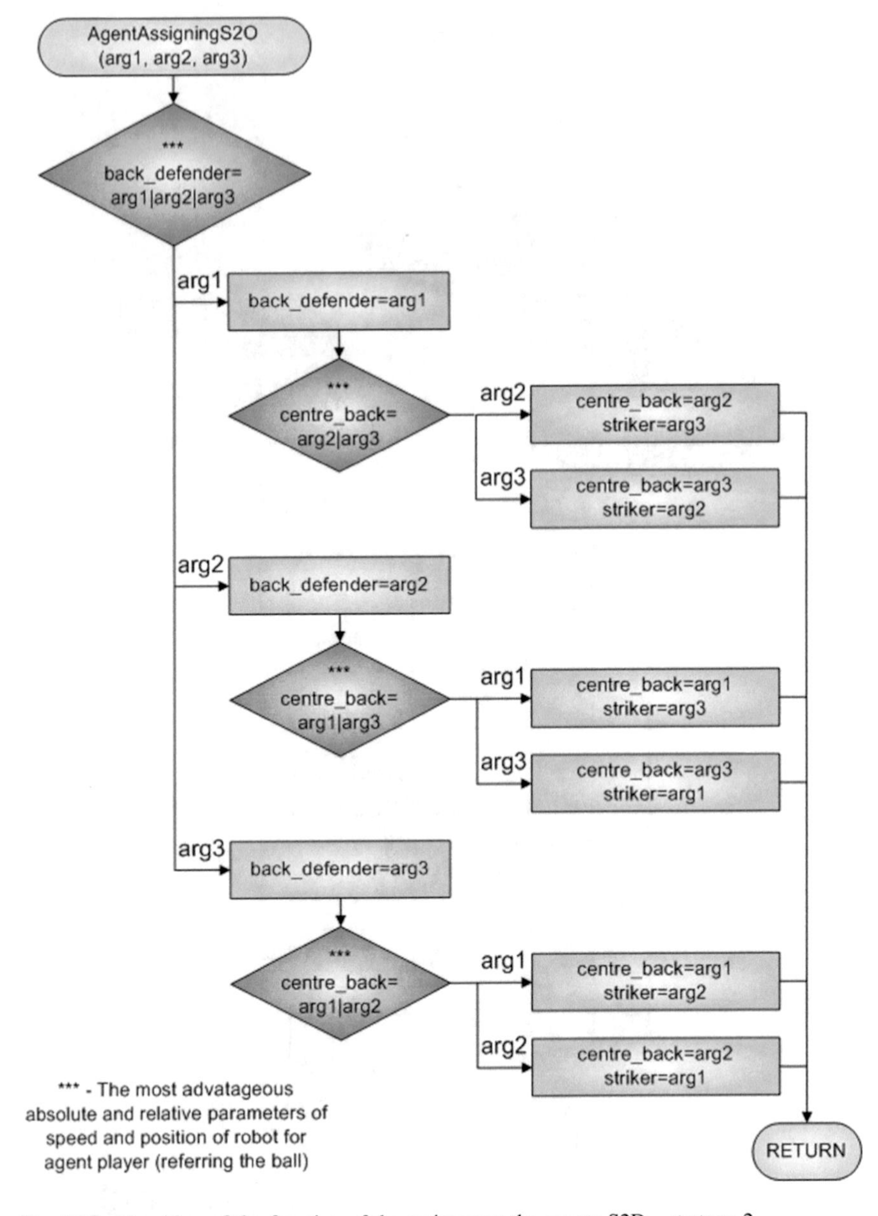

Fig. 9.18 Algorithm of the function of the assignment the agents S2D—strategy 2

The assignment of the robots to the elementary agents (Fig. 9.19) is performed in the following order:

1. Striker
2. Midfielder
3. Centre back
4. Goalkeeper
5. Defender

The rules of the creation of players by assigning the robots to the elementary agent players:

$$\begin{aligned}
&u = IDn_1 \Leftrightarrow \text{Sentence_of_position}(IDn_1, u); n_1 = 5, 4, 3, 2 \\
&z = IDn_2 \Leftrightarrow \text{Sentence_of_position}(IDn_2, z) \wedge z \neq u, n_2 = 5, 4, 3, 2 \\
&ov = IDn_3 \Leftrightarrow \text{Sentence_of_position}(IDn_3, ov) \wedge ov \neq (zl \vee u), n_3 = 5, 4, 3, 2 \\
&b = IDn_4 \Leftrightarrow \text{Sentence_of_position}(IDn_4, b) \wedge ((ov \vee z \vee u) \neq IDn_4, n_4 = 1, 2 \\
&o = IDn_5 \Leftrightarrow \text{Sentence_of_position}(IDn_5, o) \wedge o \neq (u \vee z \vee ov \vee b), n_5 = 5, 4, 3, 1, 2
\end{aligned} \tag{9.16}$$

Sentence_of_position = robot x has the most advanced absolute and relative (referring to the ball) positional and speed parameters for the elementary agent a.

The master prefers the assignment of the striker to the midfield and defender. The defender may be assigned to any robot. The defender is assigned as the last one and in such a manner, that the remaining robot is assigned to the defender. The part of algorithm of the agent master responsible for the assignment of the robots to the elementary agents is shown in the Figs. 9.20 and 9.21.

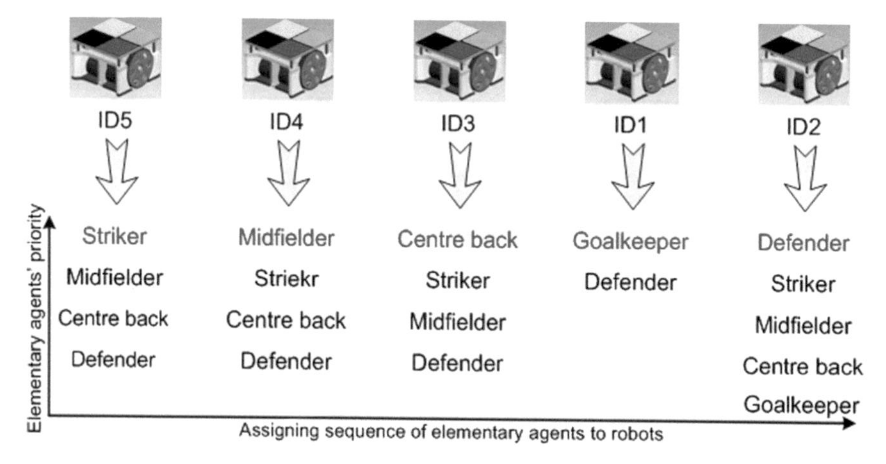

Fig. 9.19 The possibilities of the assignment of the agent players to the action midfield play—strategy 2

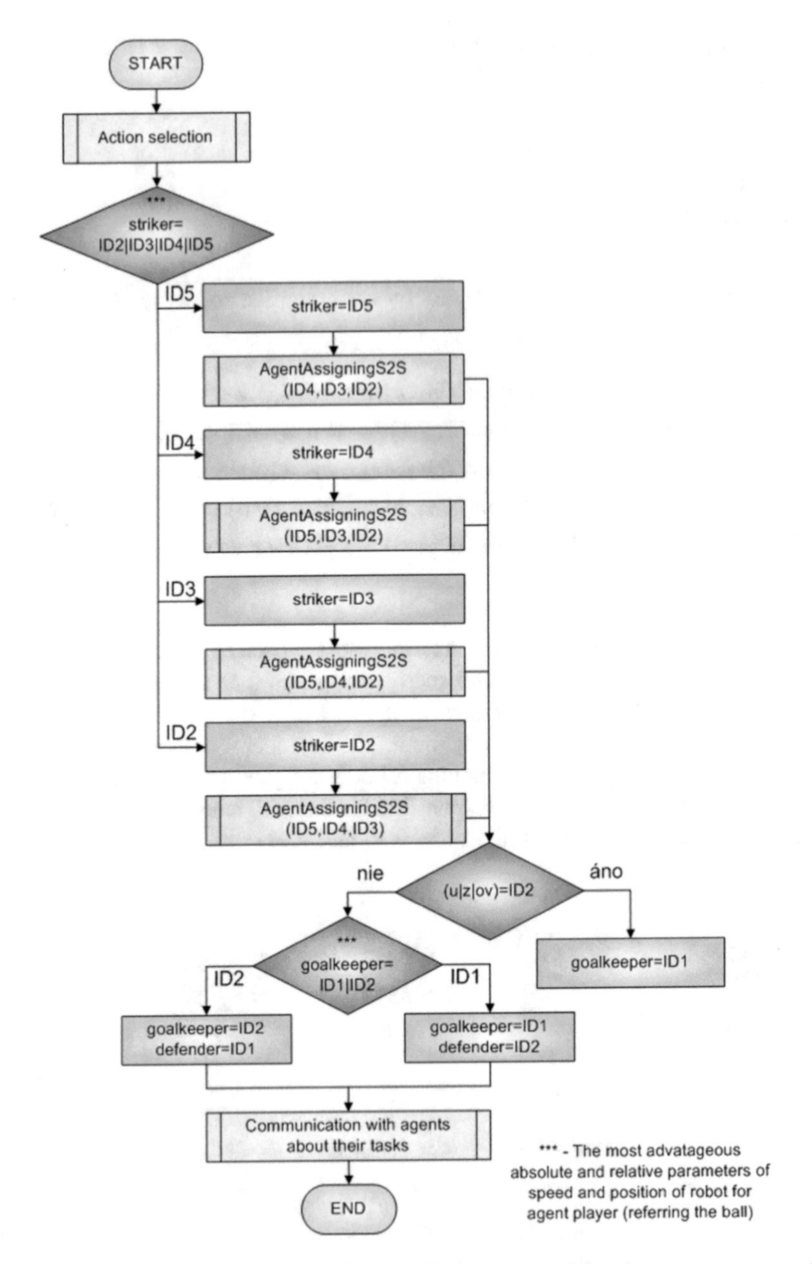

Fig. 9.20 Algorithm of the master focusing on the assignment of the elementary agents to the robots for the strategic action midfield play—strategy 2

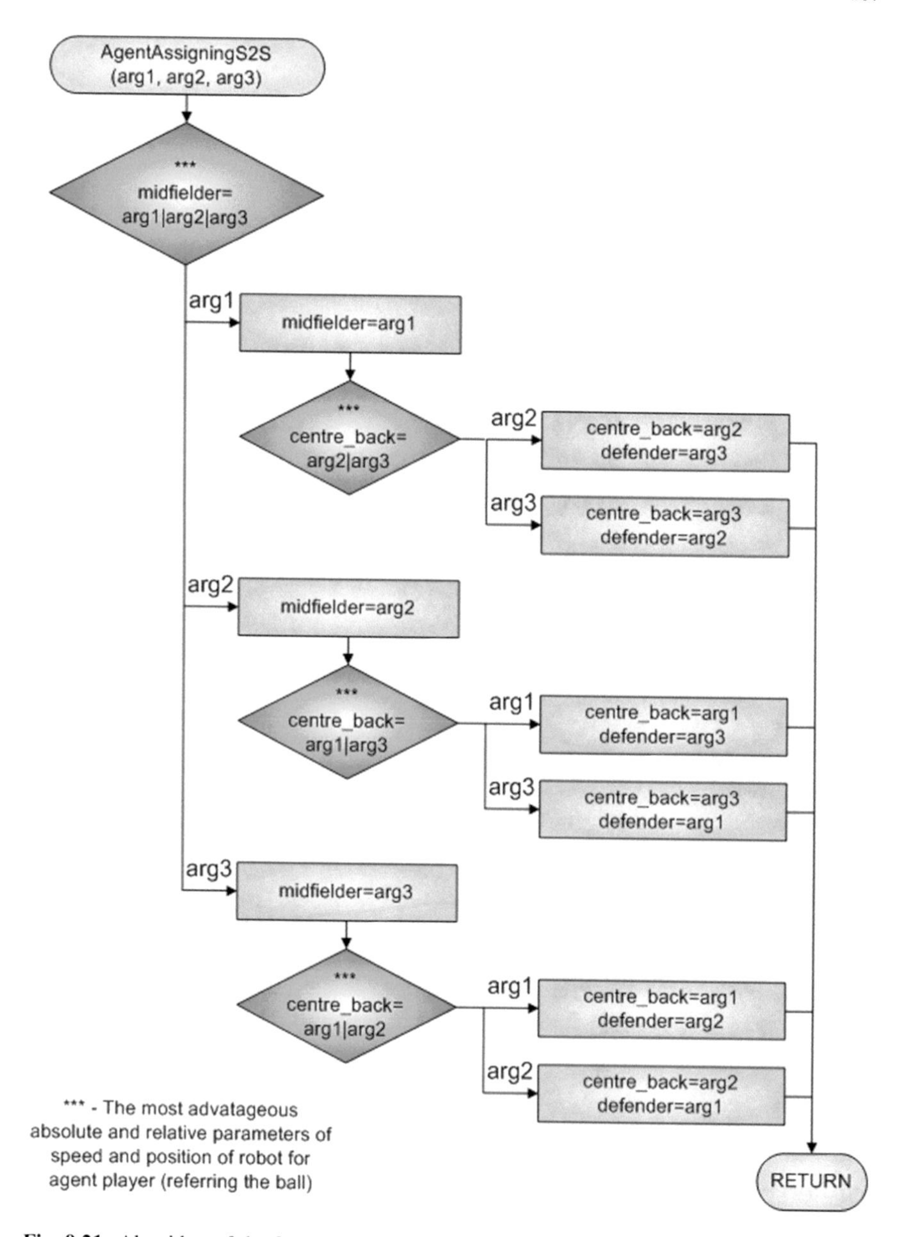

Fig. 9.21 Algorithm of the function assignment the agents in S2M—strategy 2

9.2.4 Attack

The assigned elementary agents from AH_2 are:

- Goalkeeper
- Defender
- Right forward
- Left forward
- Striker

The rules of the assignment of the robots to the elementary agents following the key:

$$
\begin{aligned}
&u \in \{ID5, ID4, ID3\} \\
&up \in (\{ID5, ID4, ID3\} - u) \\
&ul \in (\{ID5, ID4, ID3\} - u - up) \\
&b \in \{ID1, ID2\} \\
&o \in (\{ID1, ID2\} - b)
\end{aligned}
\tag{9.17}
$$

The assignment of the robots to the elementary agents (Fig. 9.22) s performed in the following order:

1. Striker
2. Right forward
3. Left forward
4. Goalkeeper
5. Defender

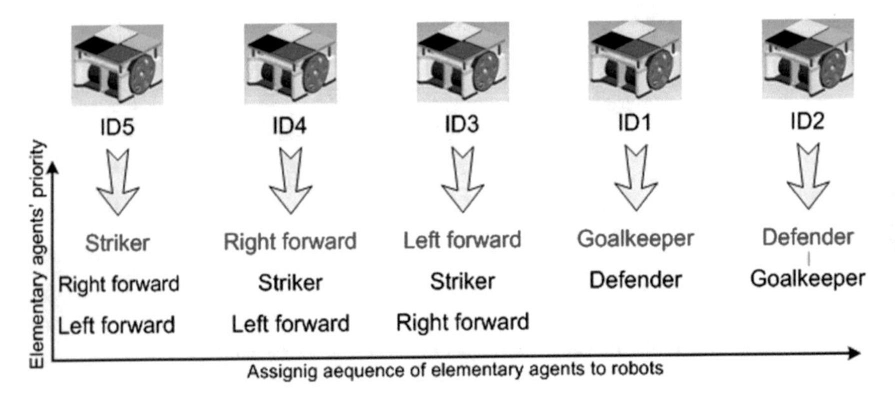

Fig. 9.22 The possibilities of the assignment of the agent players to the action attack

The rules of the creation of players by assigning the robots to the elementary agent players:

$$
\begin{aligned}
& u = IDn_1 \Leftrightarrow \text{Sentence\_of\_position}(IDn_1, u); n_1 = 5, 4, 3 \\
& up = IDn_2 \Leftrightarrow \text{Sentence\_of\_position}(IDn_2, up) \wedge up \neq u, n_2 = 5, 4, 3 \\
& ul = IDn_3 \Leftrightarrow \text{Sentence\_of\_position}(IDn_3, ul) \wedge ul \neq (u \vee up), n_3 = 5, 4, 3 \\
& b = IDn_4 \Leftrightarrow \text{Sentence\_of\_position}(IDn_4, b), n_4 = 1, 2 \\
& o = IDn_5 \Leftrightarrow \text{Sentence\_of\_position}(IDn_5, o) \wedge o \neq b, n_5 = 1, 2
\end{aligned}
\tag{9.18}
$$

Sentence_of_position (x, a) = robot x has the most advanced absolute and relative (referring to the ball) positional and speed parameters for the elementary agent a.

Figure 9.23 depicts the part of algorithm of the agent master responsible for the assignment of the robots to the elementary agent players.

9.2.5 *Total Attack*

The assigned elementary agents from AH_2 set are:

- Goalkeeper
- Midfielder
- Right forward
- Left forward
- Striker

The rules of the assignment of the robots to the elementary agents following the key:

$$
\begin{aligned}
& u \in \{ID5, ID4, ID3, ID2\} \\
& up \in (\{ID5, ID4, ID3, ID2\} - u) \\
& ul \in (\{ID5, ID4, ID3, ID2\} - u - up) \\
& o \in (\{ID5, ID4, ID3, ID2\} - u - up - ul) \\
& b = ID1
\end{aligned}
\tag{9.19}
$$

The assignment of the elementary agents (Fig. 9.24) is performed in the following order:

1. Striker
2. Right forward
3. Left forward
4. Goalkeeper
5. Defender

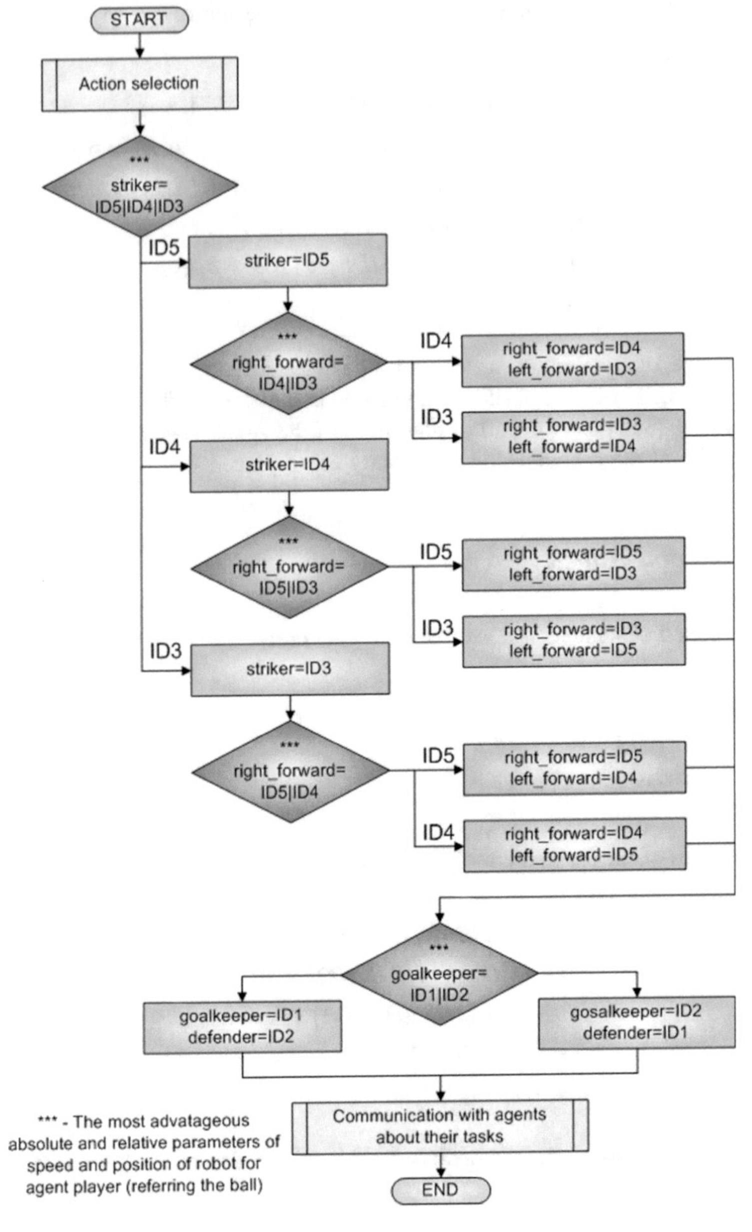

Fig. 9.23 Algorithm of the master focusing on the assignment of the elementary agents to the robots for strategic action attack—strategy 2

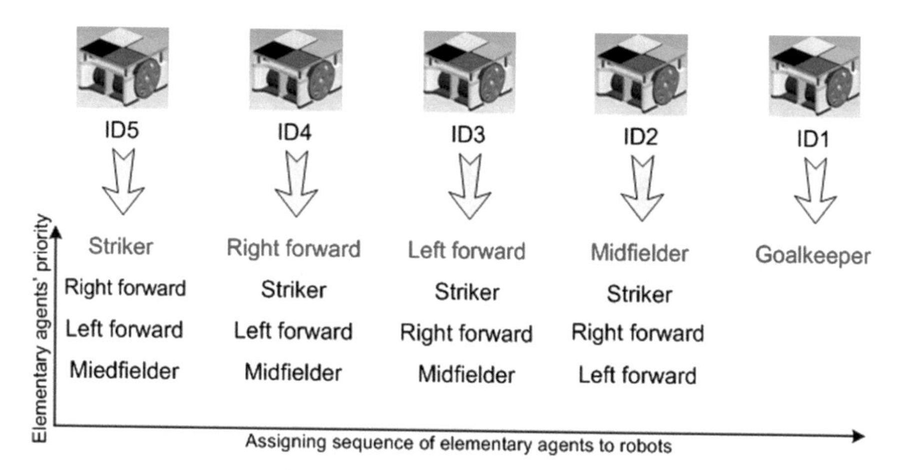

Fig. 9.24 The possibilities of the assignment of the agent players to the action attack—strategy 2

The rules of the creation of players by assigning the robots to the elementary agent players:

$$\begin{aligned}
&u = IDn_1 \Leftrightarrow \text{Sentence_of_position}(IDn_1, u); n_1 = 5,4,3,2 \\
&up = IDn_2 \Leftrightarrow \text{Sentence_of_position}(IDn_2, up) \wedge up \neq u, n_2 = 5,4,3,2 \\
&ul = IDn_3 \Leftrightarrow \text{Sentence_of_position}(IDn_3, ul) \wedge ul \neq (u \vee up), n_3 = 5,4,3,2 \\
&z = IDn_4 \Leftrightarrow \text{Sentence_of_position}(IDn_4, z) \wedge z \neq (u \vee up \vee ul), n_4 = 1,2 \\
&b = ID1
\end{aligned} \tag{9.20}$$

Sentence_of_position (x, a) = robot x has the most advanced absolute and relative (referring to the ball) positional and speed parameters for the elementary agent a.

The part of algorithm of the agent master responsible for the assignment of the robots to the elementary agents is depicted in the Figs. 9.25 and 9.26.

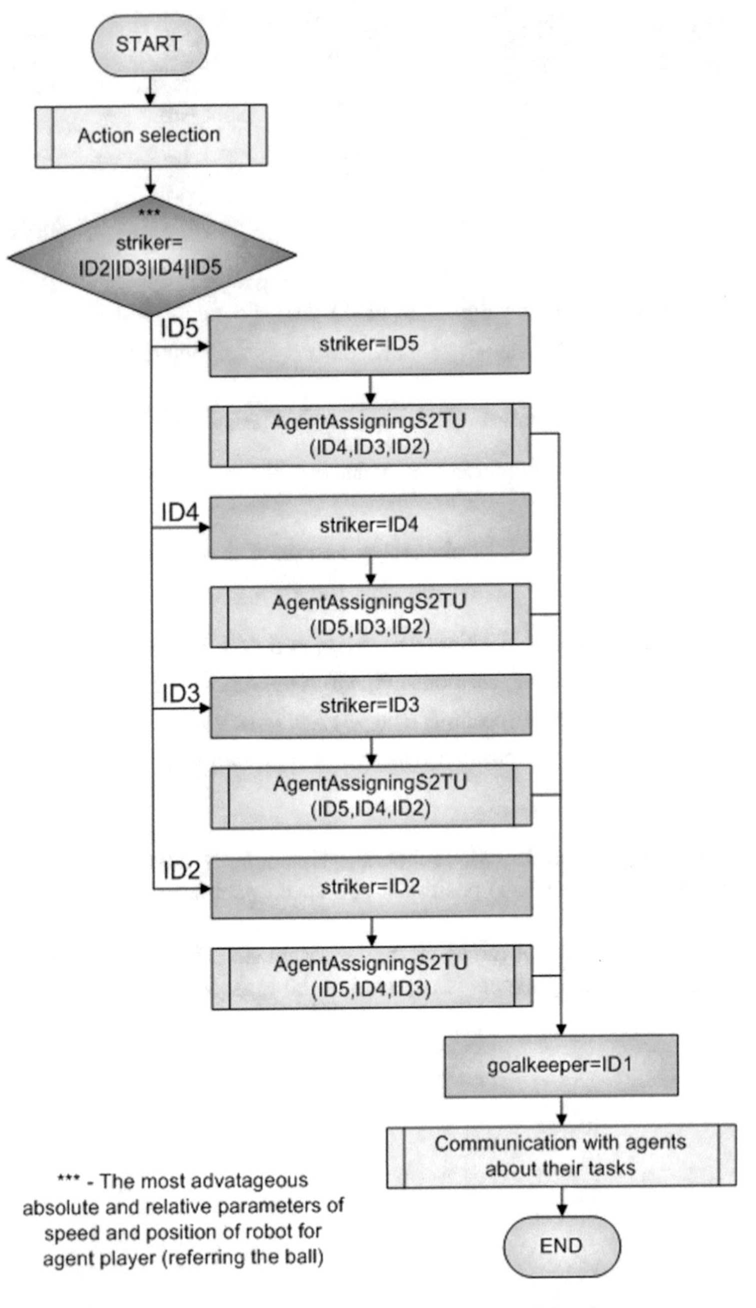

Fig. 9.25 Algorithm of the master focusing on the assignment of the elementary agents to the robots for the strategic action total attack—strategy 2

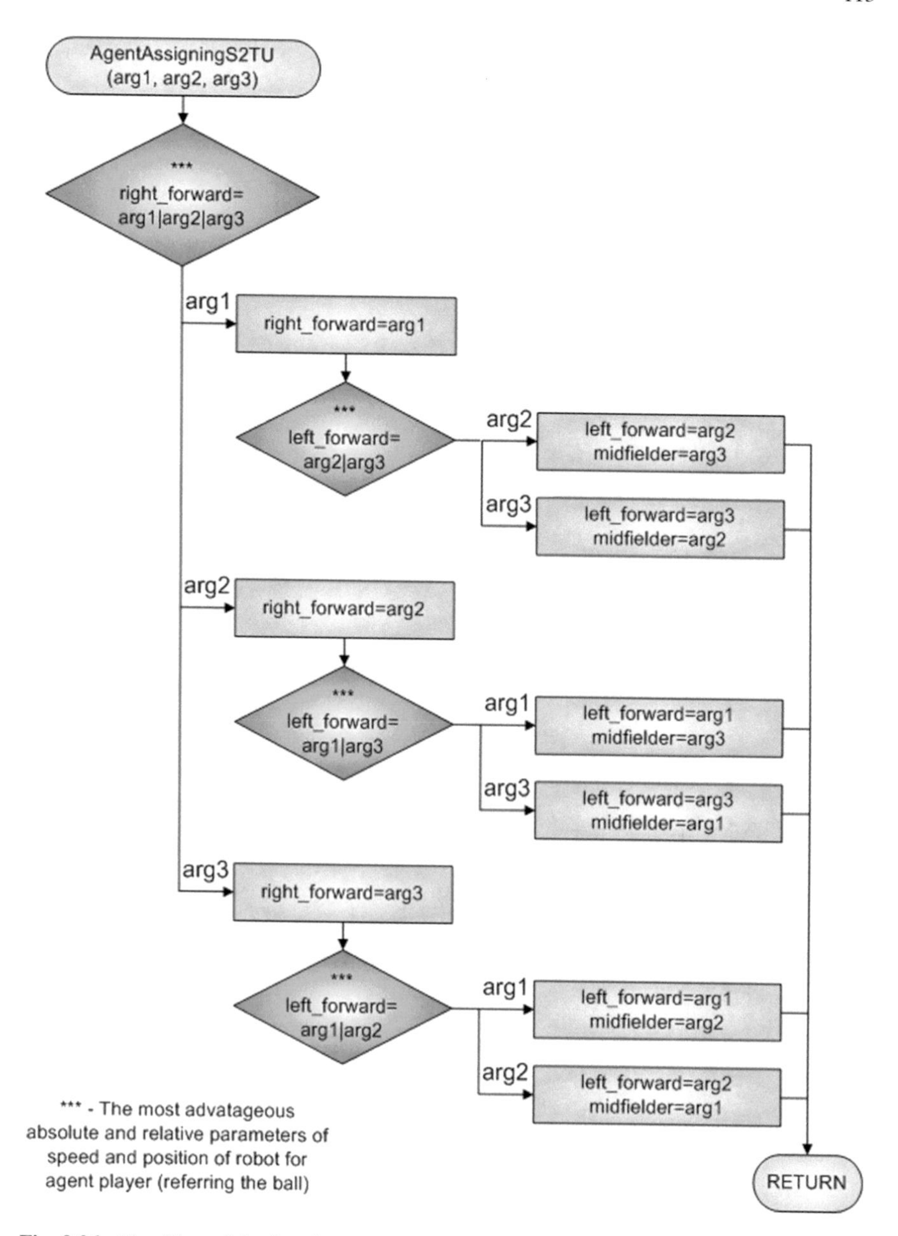

Fig. 9.26 Algorithm of the function Assignment the agents S2TD

Chapter 10
The Driver and Its GUI (Graphical User Interface)

Software for the applied control of the robosoccer team was written in Borland Builder language C++ 6.0. The principle the whole system control is based on is created by the classes. The basic agents of the system were created by the appropriate interaction of these classes. Auxiliary classes were created for the correct system function in order to set the attributes of image, image process. Apart from these, the other classes were created as well that provide the set the variables and other parameters needed for the system setting, Graphical user interface (GUI) between the computer and the user is created by means of graphical classes of the library VCL. The basic priorities at the design of the graphical interface were transparency and simplicity of the basic operating and information controls. The operator must have possibility before game to transparently set all the parameters required. The operator also must be able to access the strategy changes very quickly throughout the game during the interruptions by the referee without the selection and decreasing the number of breaks (4 possible breaks lasting for 2 min).

10.1 Description of the Program Windows

After the program has been activated, and the basic camera properties have been chosen the main program window is activated (Fig. 10.1).

The main software window is composed of the following areas:

(a) *Main menu* is composed of the items Camera, about_program (Fig. 10.2) and the End. In the camera item there are forms for the basic setting of the camera system (brightness, signal enhancement, white balance, shutter time, …)
(b) *The field (live)* is composed of the image being taken from the camera
(c) The bookmark called "colour detection" serves for the switching on the visual check of colour setting (Fig. 10.3).

M. Hajduk et al., *Cognitive Multi-agent Systems*, Studies in Systems, Decision and Control 138, https://doi.org/10.1007/978-3-319-93687-1_10

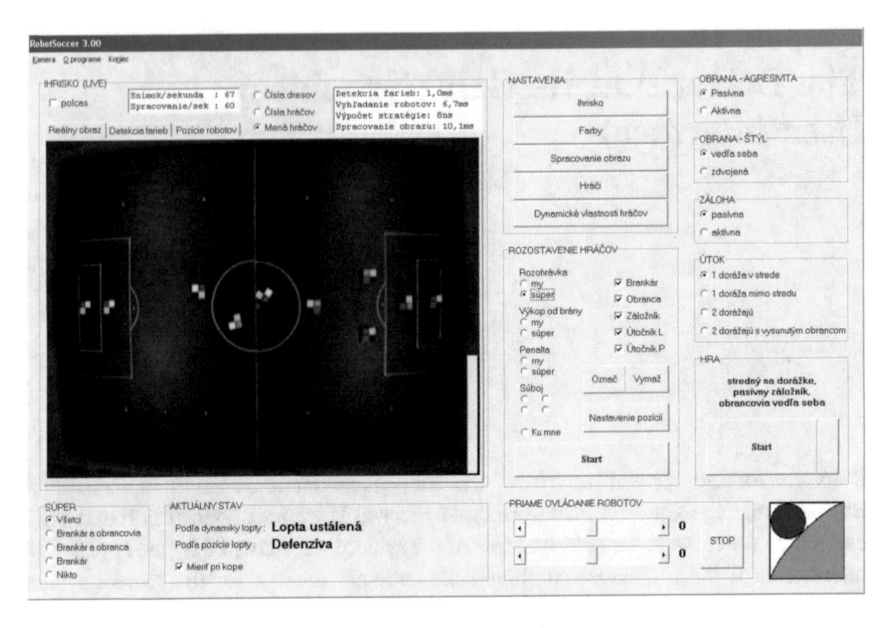

Fig. 10.1 Main window of graphic interface of control software

Fig. 10.2 Form About_program

The bookmark "robots' position" allows for the visual control of the evaluation of current situation on the field (Fig. 10.4). It is a drivers' look at the field after the image has been processed and the individual objects have been identified. This look is given to the individual agents as their inner representation of the environment.

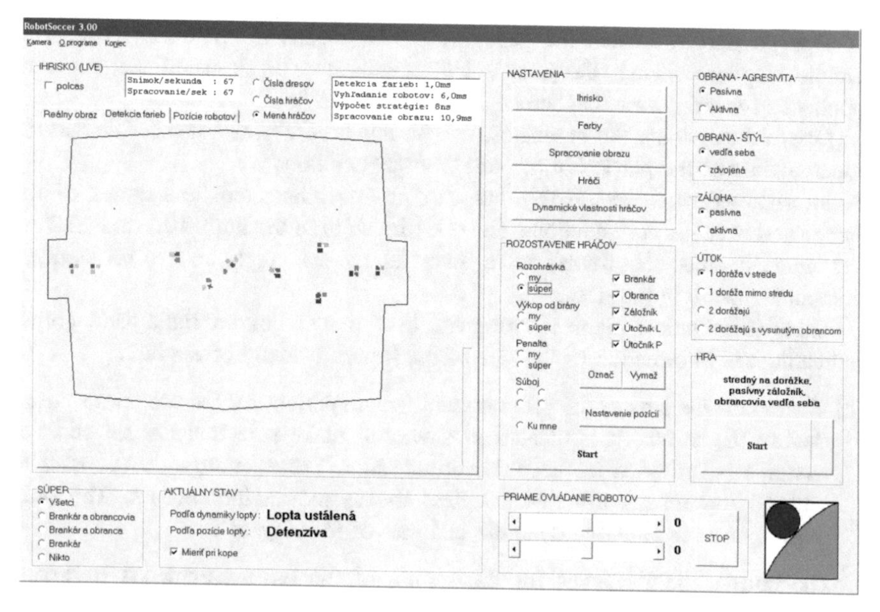

Fig. 10.3 Colour detection

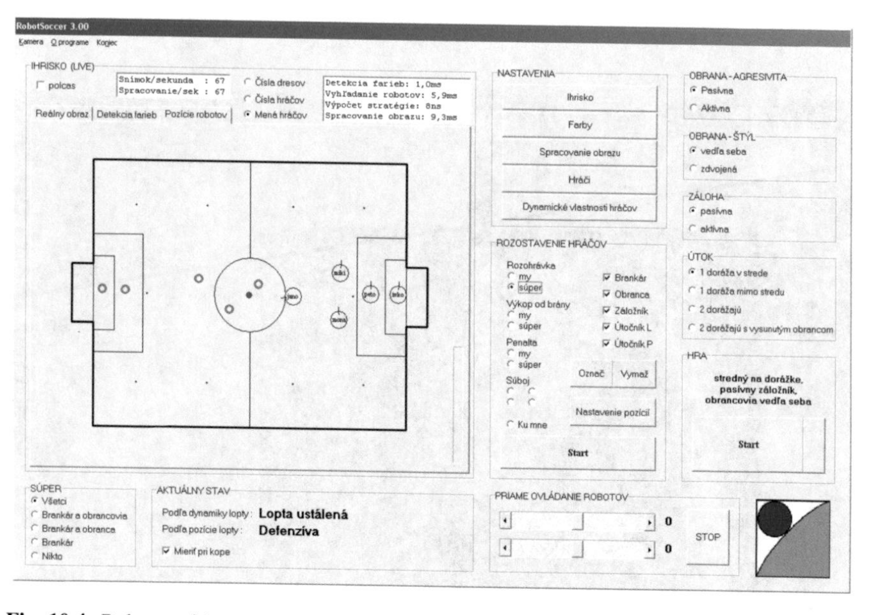

Fig. 10.4 Robot positions

It is possible to switch the game's sides by means of "half-time" cell. This simple step can switch the team's side during the break using mathematical apparatus for the mirror line across the centre of the field.

One of the options: "kit numbers", "player numbers", "player names" defines the identification of the players in the cell "robots' positions".

Information windows display the current time characteristic features of the image and strategic computations. Note: Information in the Figs. 10.1 and 10.2 are not relevant since the frames were taken in the resting time, i.e. no strategic computations are in progress.

There is graphical element "progress bar" in the bottom right hand corner indicating the information of the size of the instant velocity of the ball.

(d) *Settings* is the area of the program which is important for the program settings before the match. It is created in a way to make sure that the selection of settings is provided in the order downwards. These settings do not tend to change throughout the match unless the system failure occurs. The only exception is the button "dynamic qualities of the players".

The button "field" serves for the setting of the basic parameters of image transformation. After the form is developed (Figs. 10.5 and 10.6) the attributes are set up by means of faders to remove all the deformations in the image (fish eye, rotation, shifting, zoom, …).

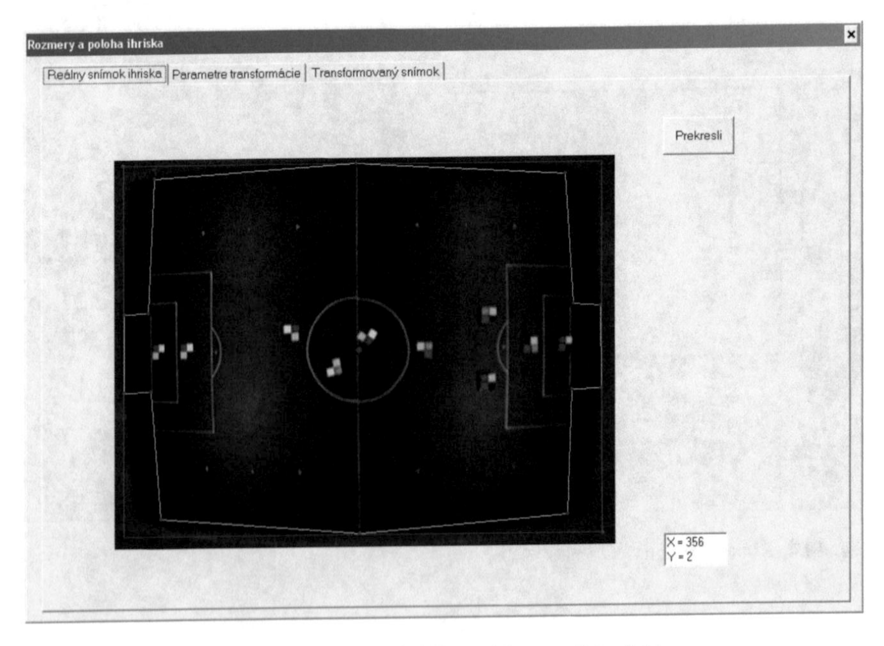

Fig. 10.5 Form "Size and position of the field"—real image of the field

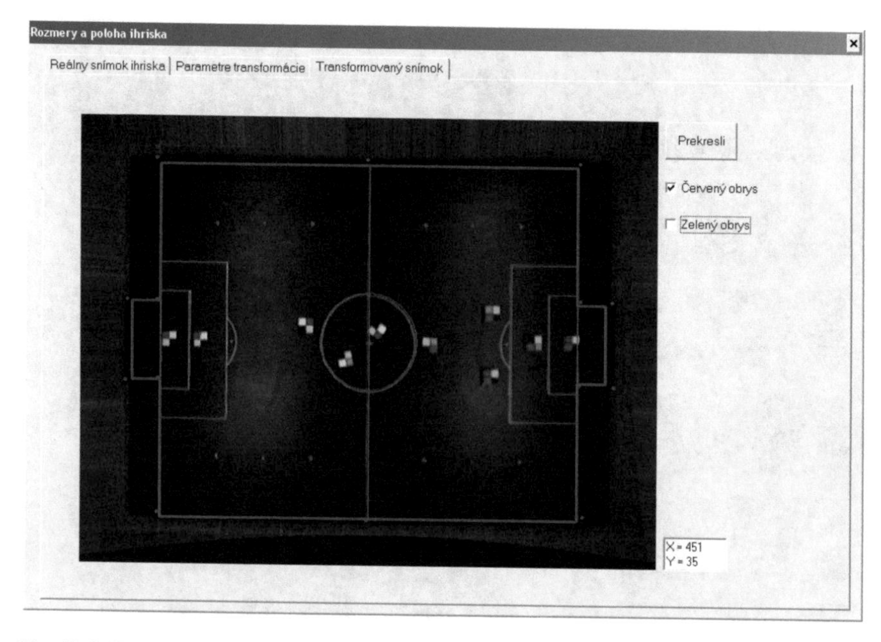

Fig. 10.6 Form "Size and position of the field"—transformed image

After the button "colours" has been pressed, the form is activated (Fig. 10.7) in which the setting of attributes of all the colours is possible: team colour (yellow and blue), basic complementary colour (green and light green), secondary complementary colour (red and pink), the ball's colour (orange). The colours might be changed and new colours might be used, but the policy of positions and colour combination must remain unchanged.

The button "image processing" activates the simple form for the setting of variables at the image processing. This setting is used with tracking the image and searching the players in 2D field of values after the colour detection has been performed. This setting can decrease the time of image processing and thus the reaction time (delay) but at the same time increase the positional and angular inaccuracy of found objects.

The form for the assignment of kits to individual players is activated after the button "players" has been pressed (Fig. 10.8). Each player is given name and its strategic position on the field. In addition, the strategic position defines the identification number of a player which is important at the process of sending data to robots.

By ticking the box "Team colour" it is defined which team colour will be chosen before the match (yellow or blue). The box "height" defines the instant height of the players. It is required for the computations related to the distortion of the player's position. Distortion of the position is higher at bigger distance of a player from the axis of a camera.

"Dynamic properties of players" button activates the form for the setting of speed and acceleration parameters of players (Fig. 10.9).

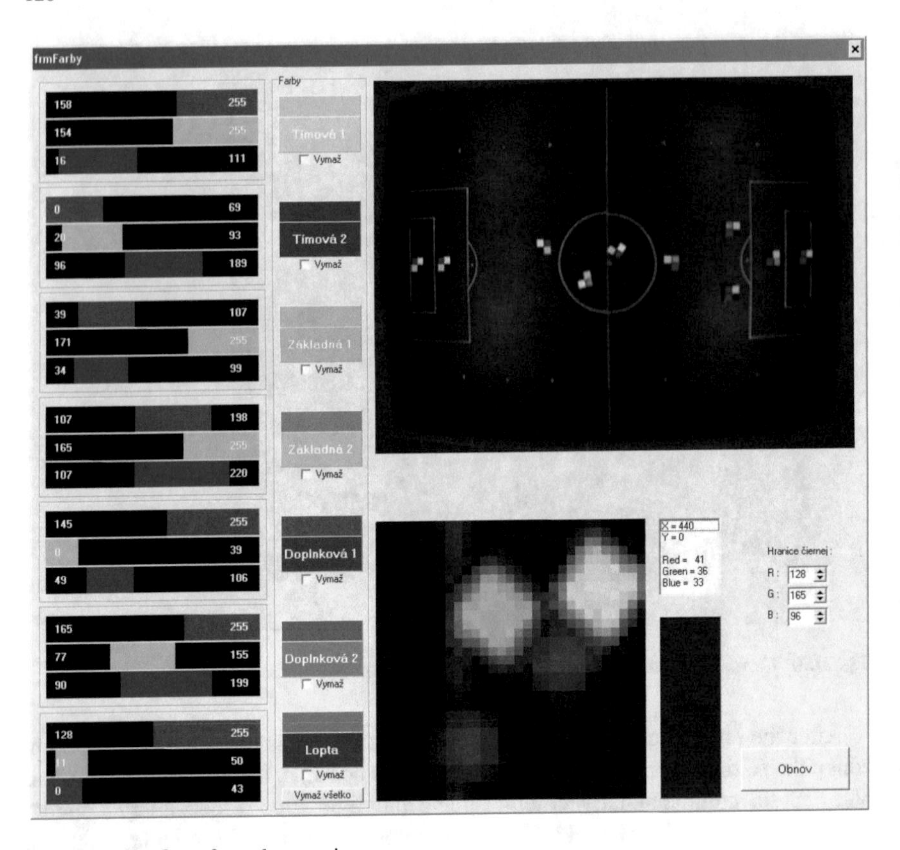

Fig. 10.7 The form for colour setting

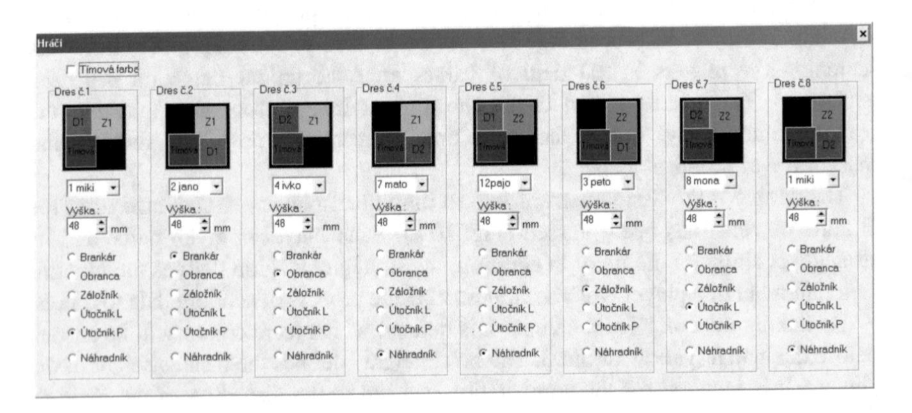

Fig. 10.8 The form “players”

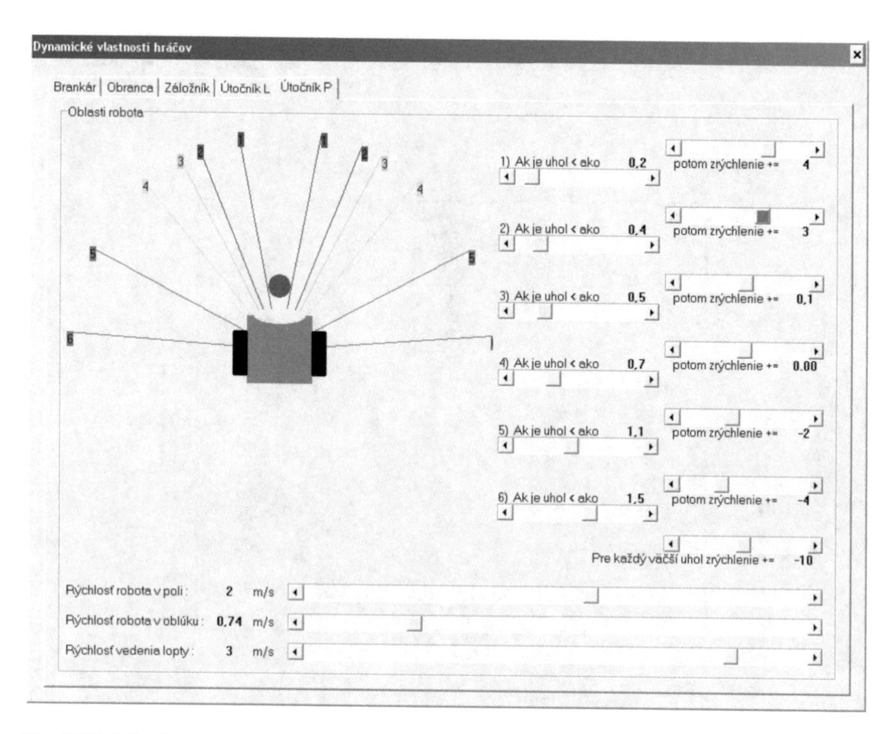

Fig. 10.9 The form "dynamic properties of players"

The setting of as many as six limit angles allows for more sensitive setting of accelerations. The angles are plotted in the left part of the form for better image of the operator. The individual speeds define the maximum speed when the basic motion actions are performed.

(e) *Positions of players* is the area used after the refereeing and decision has been made by the referee of a place and type of kick-off. After the positions have been chosen and the button "start" has been pressed, the players are navigated to pre-set positions with pre-set rotation. The navigation is fully automatic by avoiding the opponents' and own players.
Pressing "Setting positions" button activates the form in which the positional and angular parameters of players based on the type of kick-off are set up (Fig. 10.10).

This setting is designed so that the change of positions and angles was possible by means of "drop and drag" policy in order to simplify the operation and possibility to change the position of the players. During dropping and dragging of players, their current position on the field is displayed as well.

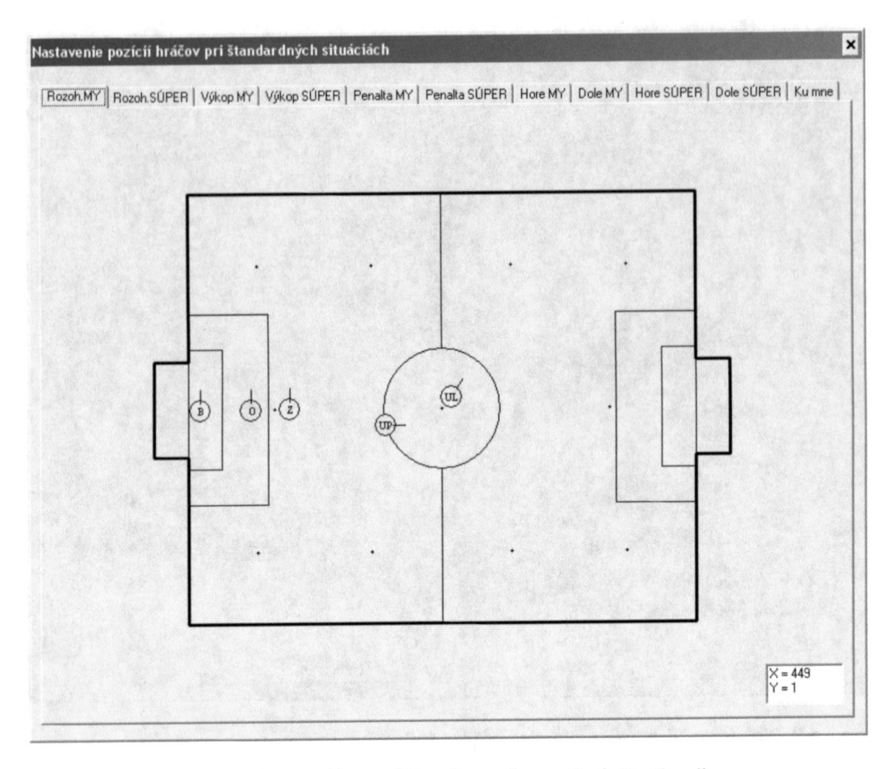

Fig. 10.10 The form "Setting positions of the players in standard situations"

(f) *The areas of setting the strategy (Defense—assertiveness and style, back-up, attack)*. It is possible to achieve 24 basic strategic positions on the field throughout the game by combining these options. All the strategic positions are listed in the Fig. 6.4.

(g) *Game* is the area given by the button "start" which starts the game. Smaller button on the right side serves for simple moving the "focus" to the button "start". In this case the game can be initiated even by pressing the key. There is check information above the buttons referring to the set up strategy.

(h) *Opponent* is the area in which it is possible to set up the way of display and incorporation of the opposing players into system in case of incorrect loading of the opponent's colour. It is also possible to affect the behaviour of players on the field since we will not allow them to see the opponent on the field and thus make decisions.

(i) *Current state* is the area composed of the current information of identifying the situation on the field by the agent captain according to the ball's direction and the ball's position on the field. This information was used mainly when the application was being developed.

Ticking box defines whether no to aim and place the ball during the process of kicking. If the option is ticked together with the option of display at least the goalkeeper of the opposing team, the player place the ball due to the convenience between the edge of the goal and a goalkeeper. In case the option is ticked, but the visibility of an opponent is not, a player place the ball to the centre of the goal. In case the option is not ticked, a player doe not place the ball, the player only stops. The ball has, in this case, random direction towards the centre of the goal because of slipping and small movement errors.

(j) *Direct robot control* is the area created due to the direct control of robot speed. The speed may be set up or directly controlled by means if arrows on the numeric keypad. Speed set up is required for trying out the right operation of remote data and robot transfer.

10.2 Program Setup Before the Game

After the camera and driver have been installed it is necessary to set up the camera according to the surrounding light (white balance, and image enhancement setup). Next, it is necessary to set up the program in the following steps:

- Field setup (removal of the deformations due to fish eye and inaccurate setup)
- Colour setup (RGB limits for used colours)
- Image processing setup
- Player positions setup (assigning colour combinations to particular robots)
- Dynamic properties of robot players setup (accelerations and speeds)
- Selection of strategy.

References

1. Agrawal, N., Jain, T.: Real time vision guided reactive strategies for micro robot soccer. A project report submitted in the partial fulfilment of the requirements for the degree of bachelor of technology. Department of Electrical Engineering, Indian Institute of Technology, Kanpur (2000)
2. Arkin, R.C.: Towards the unification of navigational planning and reactive control. In: AAAI Spring Symposium on Robot Navigation, 1989
3. Baláž, V., Sukop, M.: Multiagentnije sistemy. In: Automation: problems, ideas, decisions, 124–127. SevNTU, Sevastopol. ISBN HMC235210, 2005
4. Brooks, R.: A robust layered control system for a mobile robot. IEEE J. Robot. Autom. **RA-2** (1986)
5. Brooks, R.: Elephants don't play chess. In: Maes, P. (ed.) Designing Autonomous Agents. MIT Press (1990)
6. Brooks, R.: Intelligence without representation. Artif. Intell. (1991)
7. Brooks, R.: Intelligence without reason. In: Proceedings, IJCAI-91 (1991)
8. Cao, Y.U., Fukunaga, A.S., Kahng, A.B.: Cooperative mobile robotics: antecedents and directions. In: Arkin, B. (ed.) Autonomous Robots. Kluwer Academic Publishers, Boston (1997)
9. Connel, J.H.: SSS: a hybrid architecture applied to robot navigation. In: IEEE International Conference on Robotics and Automation, Nice, France, 1991
10. Csontó, J., Sabol, T.: Umeľinteligencia, Rektorát TU Košice (1991)
11. Dorner, J.: Multiagentové systémy zamerané na riadenie mobilných robotických systémov. In: AT&P Journal 12/2007, 94–97, 2007
12. Dowell, M.L.: Learning in multiagent systems. Submitted in parial fulfillment of the requirements for the Ph.D. in the Department of Electrical and Computer Engineering, College of Engineering, University of South Carolina, 1995
13. Drábek, O., Taufer, I., Seidl, P.: Umělé neuronové sítě—teorie a aplikace (3). Chemagazín 1 (XVI), 12–14. ISSN 1210-7409 (2006)
14. Dudek, G., Jenkin, M., Milios, E., Wilkes, D.: A taxonomy for swarm robots. In IEEE/RSJIROS, 1993
15. Franklin, S., Graesser, A.: Is it agent, or just a program?: A taxonomy for autonomous agents. In: Proceeding of the Third International Workshop on Agent Theories, Architectures and Languages, Verlag, 1997
16. Gat, E.: On three-layer architectures. In: Kortenkamp, D., Bonnasso, R.P., Murphy, R. (eds.) Artificial Intelligence and Mobile Robotics. AAAI Press (1998)

M. Hajduk et al., *Cognitive Multi-agent Systems*, Studies in Systems, Decision and Control 138, https://doi.org/10.1007/978-3-319-93687-1

17. Green, S., Hurst, L., Nagle, B., Cunningham, P., Somers, F.A., Evans R.: Software agents: a review. Intelligent Agent Group (Department of Computer Science, Trinity College, Dublin), Report, 1997
18. Novak, G., Seyr, M.: Simple path planning algorithm for two-wheeled differentially driven (2WDD) soccer robots. Wises **91–102**, 2004 (2004)
19. Hajduk, M., Sukop, M.: Multiagents system with dynamic box change for MiroSot. In: Progress in Robotics: Communications in Computer and Information Science, vol. 44, pp. 287–292. Springer-Verlag, Berlin (2009). ISBN 978-3-642-03985-0, ISSN 1865-0929
20. Hajduk, M., Sukop, M.: Multi-agent robotic system—robotic soccer in category Mirosot. Appl. Mech. Mater. **186**, 12–15 (2012). ISSN 1662-7482
21. Hajduk, M.: Humanoidný robot pre robosoccer. In: Acta Mechanica Slovaca, Roč. 13, č. 2-A, 59–64. ISSN 1335-2393 (2009)
22. Hajduk, M.: Multiagentné systémy v mobilnej robotike. In: Strojné inžinierstvo, 756–760. STU, Bratislava, ISBN 8022723142 (2005)
23. Hayes-Roth, B.: An architecture for adaptive intelligent systems. Artif. Intell. Spec. Issue Agents Interactivity (1995)
24. Jurišica, L., Murár, R.: Reaktívne riadenie mobilného robota. In AT&P Journal, vol. 5 (2003)
25. Kostelník, P.: Multiagentové systémy pre riadenie mobilných robotov. Písomná práca k dizertačnej skúške (2002)
26. Kubík, A.: Inteligentní agenty. Computer Press, Brno (2004)
27. Maes, P.: Artificial life meets entertainment: life like autonomous agents. Commun. ACM (1995)
28. Madarász, L.: Metodika situačného riadenia a jej aplikácie, 212 p. Univerzity Press, Elfa, TU Košice, ISBN 80-88786-66-5 (1997)
29. Marík, V.: Umelá inteligence I, Centa spol. s r.o., Brno, 1993
30. Marík, V.: Umelá inteligence 2, Centa spol. s r.o., Brno, 1997
31. Mataric, M.J.: Issues and approaches in the design of collective autonomous agents. Robot. Auton. Syst. (1995)
32. Mataric, M.J.: Behavior-based control: examples from navigation, learning, and group behavior. J. Exp. Theor. Artif. Intell. **9** (1997)
33. Mataric, M.J.: Behavior-based robotics. In: Wilson, R.A., Keil, F.C. (eds.) The MIT encyclopedia of cognitive sciences. MIT Press (1999)
34. Nicolescu, M., Mataric, M.J.: Extending behavior-based systems capabilities using an abstract behavior representation. In: Working Notes of the AAAI Fall Symposium on Parallel Cognition, North Falmouth, MA, 2000
35. Nwana, H.S.: Software agents: an overview. Knowl. Eng. Rev. (1996)
36. Nwana, H.S., Lee, L., Jennings, N.R.: Coordination in software agent systems. British Telecommun. Technol. J. (1996)
37. Parker, L.E.: Heterogenous multi-robot cooperation. Ph.D. thesis, MIT, EESC, 1994
38. Russel, S.J., Norvig, P.: Artificial Intelligence: A Modern Approach. Prentice Hall, Englewood Cliffs, NJ (1995)
39. Saloky, T.: Aplikácie techník strojového učenia. ELFA Košice, 1998
40. Schôn, D.: Agentovo orientované systémy. Písomná práca k dizertačnej skúške, 1998
41. Sinčák, P., Andrejková, G.: Neurónové siete I.diel, Elfa, Košice, 1996
42. Sinčák, P., Andrejková, G.: Neurónové siete 2.diel, Elfa, Košice, 1996
43. Stone, P., Veloso, M.: Multiagent systems: a survey from a machine learning perspective. Auton. Robot. **8** (2000)
44. Sukop, M.: Prenos informácií medzi riadiacim počítačom a hráčmi v aplikácii robotického futbalu. In: ROBTEP 2011: Automatizácia/Robotika v praxi, zborník príspevkov z 11. Medzinárodnej konferencie, 209–212, 16 Dec 2011, Košice, TU, SjF, ISBN 978-80-553-0846-3, 2011

45. Sukop, M.: Spracovanie obrazu a detekcia objektov v aplikácii robotického futbalu. In: ROBTEP 2011: Automatizácia/Robotika v praxi, zborník príspevkov z 11. Medzinárodnej konferencie, 16 Dec 2011, 213–216, Košice, TU, SjF, ISBN 978-80-553-0846-3, 2011
46. Sukop, M.: Information transfer between master computer and players in application of robot soccer. In: International Scientific Herald, vol. 3, no. 2, pp. 197–202. ISSN 2218-5348, 2012
47. Sukop, M.: Image processing and object founding in the robot soccer application. In: International Scientific Herald, vol. 3, no. 2, pp. 203–206. ISSN 2218-5348 (2012)
48. Sukop, M., Svetlík, J.: Communication mobile robots with PC by HF (high frequency) modules = Komunikácia mobilných robotov s PC pomocov VF (vysokofrekvenčných) modulov. In: ROBTEP, 371–376, TU, SjF, Košice. ISBN 8070998261 (2002)
49. Svetlík, J., Sukop, M.: Using of multiagent systems in robosoccer. In: Acta Mechanica Slovaca, Roč. 11, č.2-A/2007, 157–160. ISSN 1335-2393 (2007)
50. Svetlík, J.: Konštrukčné detaily robotického futbalistu 1. 2011, Transfer inovácií, č. 19, 87–90. ISSN 1337-7094 (2011)
51. Svetlík, J.: Vplyv charakteristík hardvérovej časti na dynamické vlastnosti robotického futbalistu. In: Acta Mechanica Slovaca, Roč. 12, č.2-A, 613–616. ISSN 1335-2393 (2012)
52. Woldridge, M., Jennings, N.R.: Towards a theory of cooperative problem solving. In Proceedings of the Sixth European Workshop of Modeling Autonomous Agents and Multi-Agent World, 1994
53. Wooldridge, M., Jennings, N.R.: Agent theories, architectures, and languages: a survey. In: Wooldridge, Jennings (eds.) Intelligent Agents, Verlag, Berlin, 1995
54. Wooldridge, M., Jennings, N.R.: Intelligent agents: theory and practice. Knowl. Eng. Rev. (1995)
55. Woldridge, M., Jennings, N.R.: The cooperative problem solving process. J. Logic Comput. (1999)
56. http://golfet.udg.edu/~busquets/pubs/papers/IbarraCCIA06.pdf
57. http://ls1-www.cs.uni-dortmund.de/~weiss/An%20Examplary%20Robot%20Soccer%20Vision%20System.pdf
58. http://neuron.tuke.sk/alumni/source/ui/34/beresova_bortak.pdf
59. http://osl.cs.uiuc.edu/docs/PhD-thesis-TR/PhD-thesis-TR-nov.pdf
60. http://s.i-techonline.com/Book/Robotic-Soccer/ISBN978-3-902613-21-9-rs05.pdf
61. http://www.alliedvisiontec.com/emea/home.html
62. http://www.ais.uni-bonn.de/humanoidsoccer/
63. http://www.telegraph.co.uk/technology/picture-galleries/5734786/Robots-warm-up-for-the-Robocup-Football-World-Cup-2009.html
64. www.art-brno.cz
65. www.atmel.com
66. www.faulhaber.com
67. www.fira.net
68. www.ihrt.tuwien.ac.at/robotsoccer/robotsoccer/verweiseenglish.htm
69. www.iitk.ac.in/robotics/proiectlists/soccer/microrobot
70. www.kamery.sk
71. www.mas.com
72. www.robocup.org
73. www.robohemia.cz
74. www.robosoccer.at
75. www.robosoccer.nl
76. www.robotika.cz
77. www.roznovskastredni.cz/dwnl/pel2007/06/Kosnar.pdf
78. www.tinyphoon.com/rainbow/_tinyphoon/Documents/CLAWAR_EURON_DecisionMaking.pdf
79. http://www.cs.cmu.edu/~robosoccer/main/
80. AVR221: Discrete PID controller

Printed in the United States
By Bookmasters